EVERYTHING IS SOMEWHERE

EVERYTHING

IS

SOMEWHERE

THE GEOGRAPHY QUIZ BOOK

Jack McClintock and David Helgren, Ph.D.

QUILL
William Morrow
New York

Library of Congress Cataloging-in-Publication Data

McClintock, Jack.
  Everything is somewhere.

  1. Geography—Miscellanea.   I. Helgren, David M.,
1947–   II. Title.
  G131.M35   1986   910   86-580
ISBN 0-688-05873-6 (pbk.)

Printed in the United States of America

9 10

BOOK DESIGN BY RICHARD ORIOLO

CONTENTS

IV: NORTH AMERICA

V: GEOGRAPHY OF THE MIND

FOR NORA AND ADRII

EVERYTHING IS SOMEWHERE

PREFACE

When David Helgren handed out a thirty-question quiz to the 128 students in his introductory geography classes at the University of Miami, he learned how little many of us know about our world. Almost a third could not locate the Pacific Ocean, more than a third did not know where France was, and a fifth could not locate New York City. About 7 percent, sitting in a classroom in Miami, Florida, with palm trees outside the door and a pencil in their hands, could not point out Miami on the map on their desks.

The quiz was not a scientific sampling and proved only that these particular students did not know much about the shape of their world. But the fact is, the same was true all across the country, as the press attention that came Helgren's way attested.

Helgren circulated his findings to colleagues on the faculty, several of whom telephoned him immediately to inquire, "Where's the Sahel?" After that, word began to spread. A reporter interviewed him for the student newspaper, the *Hurricane,* and a small story was published. Editors at the *Miami Herald* and *Miami News* saw the *Hurricane* article, sent reporters to talk with Helgren, and stories went out on the AP and UPI wires. Hundreds of newspapers reported on the quiz, and more than fifty of them published editorials stressing the need to teach more geography in American schools.

"THE LOST GENERATION," said headlines, were "PUTTING IGNORANCE ON THE MAP." A professor at Penn State lamented, "People in the U.S. literally don't know where they are."

Helgren found himself on *Good Morning, America.* "This is not a University of Miami phenomenon," he said. "This is a nationwide thing. UM students come here from all over. It reflects the lack of basic geographical training in grade schools and high schools."

Americans do not know much geography. Maybe because we are rich and powerful, and the rest of the world knows who and where we are, we don't bother to learn.

Of course this is arrogant, absurd, and dangerous. It also denies us a lot of fun.

Helgren got hundreds of letters and calls. Some of his correspondents wanted to use the quiz's sad results to encourage more teaching of geography. Many of them wanted to take the quiz themselves, just to see how they would do. Three years later, Helgren still gets letters and calls about his quiz. They suggest that some national nerve had been struck.

We in the United States—who call ourselves "Americans" even though every person who lives in Canada or Brazil is equally an American—are a paradoxical people. Because we are powerful and generous and meddlesome, we are internationalists. Yet we are astoundingly ignorant of the world. An average child in the Netherlands probably knows more geography than any dozen "American" adults. And at the same time, perhaps because our natural curiosity has not been satisfied in school, we want to know more. Helgren's mail proved that.

He wanted to do something about it. And that is what this book is about.

It is a geography quiz—much bigger, better, more detailed, and a lot more fun than Helgren's original quiz.

As you take it, you'll be amazed at how much geography you already know.

You will also be surprised—pleasantly, we hope—at some of the geography you don't know.

The first few chapters sketch some background about the setting in which geography exists—that is, the earth. (*Geography* comes from the Greek word for earth: *geo.*) Other chapters continue the story, through the human imprint on the earth, to the regions of the earth and their peoples. The last of these regions is our own: North America, a place of incredible . . . well, geography. At the end is a chapter on what you might call geographical ideals, or maps of the spirit—because it is the human mind and heart that gives the world meaning.

We hope you will enjoy this look at the place where we live.

We did.

INTRODUCTION: GEOGRAPHY IS EVERYWHERE

Everything *is* somewhere. And if it's somewhere, it is part of geography. As professional geographers say when they want to poach on some other specialist's turf, "Geography is everywhere."

This puts a burden on the poor geographer, of course—what one of them drolly describes as "the terrible obligation to keep track of it all." Because geography is more than knowing the names of state capitals and recognizing the silhouette of South America. You might say geography is about the world and what it's made of: the composition, distribution, and relationships of its animals, vegetables, and minerals. Or you could say that it's the study of where things are and why they are there, that it is the science of locating and explaining. That is as close as we'll try to come in this book to defining our subject.

Under a definition like that, one could include all sorts of things, from how magma is formed to who invented noodles. We do not go quite that far, although we come close. We maintain that you can't understand the world's geography, as we define it, without a fair understanding of such matters as which country makes which beer, where romantic love is found, and how desert animals cool their hindquarters.

Magma isn't everything. On the other hand, magma is something, and it's somewhere. It is certainly part of geography, so in the following pages somewhere you'll find something about magma, whatever magma is.

If you take the whole quiz, we guarantee you'll know more geography at the end than you did before—and probably that will be a lot more than your neighbor knows. And if you're a curious person, or a person who loves travel or is interested in foreign cultures, or simply loves human beings and all their strangeness, we think you'll enjoy the process.

Geographers usually love their work. Being a geographer is fun because geographers don't have to limit themselves to studying any one thing, the way geologists or economists or anthropologists do. They can study *everything*. "You just study what you enjoy—and call it geography," as one geographer says. For instance, here are some things geographers are doing these days:

- Choosing the locations for new furniture stores
- Pondering the conversion of Mark Twain's Hannibal, Missouri, into a tourist trap
- Modeling the climate of the last ice age
- Studying the housing and service needs of a two-earner family as a problem in feminist research
- Examining the impact of cars and garages on the look of houses and neighborhoods
- Monitoring soil erosion in suburbia
- Finding out how people really respond to hurricane warnings
- Determining which states provide the most professional football players
- Reconstructing the landscapes of the Old Testament
- Using satellites to map the earth's surface

Most geographers specialize. Some study landforms, climates, and natural resources. One may study cities, another where to locate factories, another popular culture. There are geographers who are experts on a region—such as South America—and other geographers who keep track of data on the whole earth.

You might say, if you didn't mind such clichés, that geography is where it's at.

As you no doubt suspected, geography begins with maps.

The simplest—but perhaps also the most complex— are the mental maps we make and carry in our heads. We use them all the time. Even our repeatedly successful discovery of the bathroom in the dark of midnight while we are nearly asleep suggests how ingrained—and important—these mental maps are. We may not think about them, but we can't get around without them.

To make the simplest trip—across the river and into the trees, or down the block and into the shopping center—we must know *how things are* out there. We need general principles of geography and navigation, such as: If you walk straight ahead, then take a left, then take a right, you're walking straight ahead again. If you head east, then north, then right, you're heading east again. And so forth.

When you give a dinner party and provide directions to your house, you're giving a little geography lesson: You explain how to get across unfamiliar terrain—a new neighborhood—to an unknown location: your house, where the wonderful food will be.

You tell your guests a sequence of turns and intervening measures of distance, some lefts and rights, some norths and souths, some landmarks ("left at Wing Fat's Pasta Parlor and Salad Bar, south along the coast highway through three traffic lights and past the sign that says BEER—WORMS—JESUS SAVES—POODLE GROOMING, hang a right at the Seven-Eleven, and it's the sixth house on the left with the pink plastic flamingo in the yard.") And if you live in a place that is laid out sensibly, your guests arrive safely, if perhaps too early and bearing a bottle of inappropriate wine.

All this is not a dull process, though, is it? Just ponder one aspect, for instance: Why do some of your friends prefer their directions verbally, the way they are given in the example above, and others prefer them in the form of a map? We suppose this has to do with left-brain and right-brain dominance, but either way, it's mental mapping.

You use a mental map just driving to the office, or to the airport. Part of the weirdness of flying is covering such immense distances without actually *experiencing* it, so that when you disembark a few hours later, and the people are exotic, moving in and out of unusual buildings on mysterious errands on a different kind of earth—the red-pink of Marrakech, for instance, which looks the same but is very different from the red-pink of Santa Fe or Georgia—you need a brand-new mental map to become oriented.

Sometimes your mental map gets disoriented, as when you emerge from a subway and start walking, thinking you are heading downtown, only to find yourself puzzled, then stopping with a lurch, realizing you've been heading uptown all the time. And now you have to haul the entire city around 180 degrees in your mind.

Geography may begin with maps, but they are only representations, not the real thing. They help you get located, but give you no vivid sense of a place—it takes the place itself to do that. We easily recognize the differences between places we love—mountains, beaches, certain lovely cities—and places we hate— such as slums and toxic-waste dumps.

Of course it's not always easy to know where you are, especially in the United States, where every airport seems to have the same brand-name logos on the wall—Holiday Inn, American Express—and you have to examine the postcard rack to see something "local"— alligators in Miami, the Empire State Building in New York. Symbols of places. (Sometimes the signs in airports help you in funny ways: There's one in the Beijing Airport, in English, that says "Chinese Restaurant.")

The human mind makes maps easily—even unconsciously. That's lucky, because it may be very important for us to have a clear picture of the larger world—exactly how big and far away Central America is, for instance, what its products and population are, and what it is really like.

We have road maps, almanacs, atlases, navigational charts, satellite navigational tracking systems, little compasses on the dashboards of our cars. But we still have to find our way from place to place, and somehow get back to where we came from, and give directions to others.

That's life.

That's geography.

After all, where would we be without geography?

Right.

Nowhere.

THE WHOLE EARTH

It's important to see things whole, especially if you are going to go poking around among their parts, the way we will. Some say that the true beginning of the environmental movement came with our first view of Earth from the moon. That was the first time anyone had seen this planet of ours dangling in space, so lovely, so delicate, so brave. We glimpsed its vulnerability then, and maybe began to see how we are all in this together. We may be traveling with some pretty strange companions—as we shall see—but we are all going in the same direction, at the same speed, and have a lot in common to ponder. These opening questions offer a glimpse of the earth as a whole:

1. Name the native continent of these animal groups:
  a Grizzly bear, raccoon, puma
  b Kangaroo, dingo, wallaby
  c Black panther, tapir, guinea pig
  d Lion, giraffe, impala
  e Tiger, sun bear, gibbon
  f Roe deer, Barbary ape
  g Penguin, krill

2. Was Columbus first to prove the earth is not flat?

3. What is the shape of the earth?

4. Name some ways you could show that the earth is spherical.

5. What point on the earth's surface is farthest from the center of the globe?

6. Where on earth is the Texas Cowboy Club?
Luckenbach, Texas
West Berlin, Germany
Paris, Texas
Marrakech, Morocco
Paris, France

San Antonio, Texas
Bombay, India

7. How far apart are lines of latitude?

8. How far apart are lines of longitude?

9. Do lines of longitude have anything to do with measuring time?

10. **a** How many time zones are there in the continental United States?
  b How many National League baseball clubs play at home in the Central Time Zone?

11. How many time zones are there in China?

12. How many time zones are there in the USSR?

13. At what international border do you make the biggest leap in time, from zone to zone?

14. If you were to put a *tanga* on a *garota*—or took one off, for that matter—what would you have?

15. You know there is a connection between the moon

and the tides. But why, exactly, is it that the moon rises about fifty minutes later each day and the high tides come along about that much later also?

16. If you went into your backyard and started digging and dug all the way through the center of the earth to the other side, would you end up in China? Or where? How would you find out the answer?

17. What is called "the intellectual's catsup" and in what major cuisines of the world is it featured?

18. If you live in the United States, in which direction would you ordinarily fly to reach the Far East?

19. This is the first Chicken Question. There will be other Chicken Questions. As for the chicken, never mind whether it came before the egg or after the egg, or why it crosses the road—can you tell where the chicken came from in the first place?

20. In what exact direction do you travel when you go through the Panama Canal from the Caribbean Sea to the Pacific Ocean?

21. There's a lot of water on the earth, in oceans, lakes, rivers, streams, rain, clouds, and so forth. About how much of it is fresh water?
97 percent
3 percent
33 percent
79 percent

22. Here is a famous quotation from the linguist Alfred Korzybski. Add the missing word: "The _____ is not the territory."

23. Which country would you say has the highest murder rate in the world?
Japan
United States
Brazil
USSR
Costa Rica
Italy
Paraguay

24. Can you name two important cities, very far apart, that are known for their cherry blossoms? For extra credit, can you tell how they are connected?

25. To be useful, every map should contain five elements. How many can you name?

26. At Babson College, in Wellesley, Massachusetts, there is an unusual map. Do you happen to know what sort of map it is?

27. The polar regions of the earth lose more energy to space than they get from the sun—and thus they stay cold. The equatorial regions get more energy from the sun than they lose to space—and they are warmer. What happens to the surplus heat from the equatorial zone? And about where, in the Northern Hemisphere, is the dividing line between zones?
Around Madrid, Philadelphia, and Beijing
Around Havana, Guangzhou, and Muscat
Around Bogotá, Singapore, and Monrovia

28. Where is the only wildlife refuge in the world that has a subway running through it?

29. *Albedo* is a term for the percentage of energy reflected from an object. For example, the albedo of ordinary dirt is about ten, meaning that 10 percent of the energy that hits the dust is reflected back. What would you guess the earth's albedo is?
About ten
About thirty
About fifty
About seventy
About ninety

30. Do people live on the surface of the earth? Hint: This is sort of a trick question.

31. Like the rest of us, you are made up of a variety of chemical elements. One of them is carbon. The carbon in your ring finger is identical to the carbon in the diamond in the ring you may be wearing. Where did the carbon in your ring finger come from?

32. During Marco Polo's day, the most accomplished navigators and mapmakers were the Arabs, whose books of maps were worth a fortune to Western navigators. But these maps were inconvenient to Western navigators in one small way. What was this?

33. Match up which continent . . .

Is most densely populated	Africa
Has the longest river system	Europe
Has the lowest inland point	South America
Has the highest waterfall	North America
Is the world's largest island	Antarctica
Is the site of the South Pole	Asia

34. There is only a single point on the globe from which you can see, with the naked eye, both the Atlantic and Pacific oceans. Where would you say that is?

35. Garrison Keillor of *Prairie Home Companion* points out one of geography's great conundrums: Why is it that you can fly westward forever and you're always flying west, or travel eastward for as long as you like and your direction will never change—it remains easterly—but if you go north, sooner or later your direction will become south, and vice versa?

36. What is the largest country . . .
In Africa?
In South America?
In Europe?
In Asia?
In North America?

37. What is the most populous country . . .
In Africa?
In South America?
In Europe?

38. On what continent is the world's largest pyramid?

39. Where was the Land O' Goshen?

40. On what continent would you find the most violent country in the world?

41. The number of sovereign nations on each continent varies, of course. Match the continent with the number of countries:

1	North America	a	Fifty-three
2	Africa	b	Twelve
3	Europe	c	None
4	Australia and Oceania	d	Forty-one
5	South America	e	Thirty-three
6	Asia	f	Twenty-two
7	Antarctica	g	Eight

THE WHOLE EARTH

1. **a** North America
   b Australia
   c South America
   d Africa
   e Asia
   f Europe
   g Antarctica

2. People knew for centuries before Columbus that the world wasn't flat—it was mostly in backward Europe that flat-earth notions held sway. Aristotle and Pythagoras, for instance, presumed the world was round. Eratosthenes, the Greek librarian at Alexandria in Egypt, actually produced a very good estimate of the earth's circumference about 200 B.C. He had noticed on a certain day that in Syrene, a city in southern Egypt, the sun shone straight to the bottom of a deep well, that is, the sun was vertical at noon this day in Syrene. But on the same day in Alexandria, the sun was not straight overhead, but seven and a half degrees off the vertical. By estimating the distance between Alexandria and Syrene, and using trigonometry, Eratosthenes computed the earth's circumference at 24,900 miles. He was under by only 4,800 miles.

3. It certainly isn't round, as almost everyone knows. It's an oblate spheroid, a sphere compressed at the poles, like a half-deflated volleyball you sat on. It bulges some at the equator, mainly because of the centrifugal force of rotation, so that at the equator, the earth's 7,926-mile diameter is 26 miles greater than its polar diameter. That's impressive, when you think that Mount Everest is less than 6 miles high.

4. Here are four ways to show that the earth is basically round: (a) go to outer space on a rocket and look back; (b) watch objects moving away from you and see how they sink over the horizon instead of merely getting smaller; (c) observe a lunar eclipse and note that the earth's shadow is always circular as it moves across that of the moon; (d) travel from the North Pole to the equator and watch the North Star, which is directly overhead at the first spot and exactly on the horizon at the second. As you travel, keep track of the changing angle of observation of the star, and you'll see that your path has been an arc—or half a sphere. This, incidentally, is the basis of stellar navigation.

5. The farthest-out point is not Mount Everest. Because of the earth's equatorial bulge, the farthest-out spot is Mount Chimborazo in the Ecuadorian Andes. (It is 20,561 feet high.)

6. The Texas Cowboy Club is in West Berlin. Members have a clubhouse fixed up just like a nineteenth-century Abilene saloon, with a Lone Star flag and swinging doors and everything. Here, the Germans array themselves in gunfighter outfits and belly up to the bar to drink insipid American beer from longneck bottles and listen to Willie Nelson records and remember the Alamo. But don't think the Germans are alone in such silliness. There are people in France who think Jerry Lewis movies are art.

7. Lines of latitude, called parallels because they are, run east-west and measure north-south position. You see them on any globe or map. They are measured from zero to ninety degrees in both hemispheres, for

reasons that are evident when you look at them—ninety degrees, the arc between equator and pole, is all there is to a hemisphere. Lines of latitude are always about 69 miles apart. If you want to be picky, the earth's oblateness causes them to be 69.4 miles apart at the poles and 68.7 miles apart at the equator.

8. Lines of longitude are never parallel, and thus their distance apart varies. They are called meridians, and they connect the poles. Meridians are those north-south lines that measure east-west position. They are farthest apart at the equator—zero degrees latitude, where they are about sixty-nine miles distant from one another.

9. Lots. Despite newfangled and spectacularly accurate timepieces that measure time by the decay of radioactive elements, measuring days and nights—the period of earth rotation—remains fundamental. Since longitude lines measure east-west position, the relative movement of the sun across the sky can be gauged by the degrees of longitude the sun passes over. There are 360 degrees of longitude. By convention, the day is twenty-four hours long. Divide the 360 degrees by twenty-four hours and you get fifteen, and thus, ideally, each one-hour time zone is 15 degrees of longitude wide. "Ideally," because time zones are not actually 15 degrees wide. They are adjusted for political and economic convenience, so as to avoid difficulties such as having two different time zones inside one town. Time zones are numbered from the Royal Observatory at Greenwich, England, which is also where the numbers on meridians start. Washington, D.C., is on 77 degrees west longitude and is five hours earlier than the mean time at Greenwich. Washington, D.C. (and the Eastern Time Zone) is GMT −5. On the other hand, Moscow, USSR, is on 37 degrees east longitude and is three hours later than the chronometer in Greenwich. Moscow time is GMT +3.

10. a Portland, Maine, is at 70.16 degrees west longitude and Portland, Oregon, is at 123.41, a difference of 53.25 degrees. If you apply what we learned in question 8, above, you can see that the width of the continental United States is a bit over three solar hours—or three time zones.

    b Three National League baseball teams play at home in the Central Time Zone: the Chicago Cubs, the Saint Louis Cardinals, and the Houston Astros.

11. China extends all the way from about 74 degrees east longitude to 135 degrees east longitude, or about 61 degrees. It is so wide that when the sun is rising in Beijing, it is still the middle of the night in western Xinjiang Province. Thus, China should have four time

zones. But by edict all clocks in China are set to Beijing time, and there is a single time zone in the entire vast country. The Chinese are all eight hours ahead of Greenwich Mean Time.

12. The Soviet Union extends from about 23 degrees east longitude to nearly 168 degrees west longitude—about 169 degrees. Thus the USSR should have eleven time zones, and that is how many it has. When the time is 8:00 A.M. in eastern Siberia, it is 9:00 P.M. on the previous day in Moscow.

13. For this, you'll do well if you recall China's single time zone—Greenwich Mean Time (GMT) plus eight hours. The time zone of neighboring Afghanistan is GMT plus four and a half hours. Subtract Afghanistan's time from China's time and you have a three-and-a-half-hour time change. The few nomads who get to cross this border probably do not care. In most parts of the world, time zones are a mere one hour different.

14. Well, defining the terms will help clarify matters. A *tanga* is one of those amazing backless string bikinis first worn on the beaches of Brazil, probably by a *garota*. *Garota* is a useful Brazilian Portuguese word for the human female of that particular age when, half-girl and half-woman, so apparently knowing but not quite fully conscious of her own true innocence, she can stroll along the fabled beaches of Copacabana and Ipanema and, especially if wearing a *tanga*, have the most powerful effect of her life upon men of a certain age. Listen, we know this may be sexist, but that's how they feel about these things in Brazil.

15. The moon rises later each day because, as the world turns, it also revolves around the sun. Thus the moon, in making a trip around the earth, also has to travel down the earth's orbital path for a time, which amounts to fifty minutes a day, relative to an observer on earth.

16. If your backyard is in Rio de Janiero, you would end up in China if you dug straight through the earth. But not if you began digging in Westport or Seattle or Manila or Umatilla or Amarillo. Here's a method for finding your antipodal point, as this is called. Look down the street and find a neighbor with a boat, and ask him or her the latitude and longitude of your house. Or look on the topographic map of your neighborhood. Then go 180 degrees of longitude east or west from your house's longitude, to the opposite side of the globe. Then go to your latitude in the opposite hemisphere. Say your house is in Honolulu, at 21 degrees 18 minutes north and 157 degrees 50 minutes west, nice palm trees outside and a balmy Pacific breeze. Go 180 degrees east or west, whichever you like, and you'd be at 22 degrees 10 minutes east. Then go 21 degrees 18 minutes south (because you were north). Where are you? You're in the northern Kalahari Desert of Botswana. Hot?

17. By some garlic-loving intellectuals, garlic is known as the intellectual's catsup. It is used by cooks—smart and dumb, lettered and unlettered, with recipes or freehand—all over the world, thus proving that no nation has a monopoly on intelligence, or garlic. In our opinion, garlic is one of the world's two foodstuffs that you can't ever use too much of. The other is chocolate.

18. From the United States, you fly west to reach the Far East.

19. The chicken, blessed dumb fowl, was domesticated at least five thousand years ago in Southeast Asia. People all over the world have invited it to dinner ever since.

20. From northwest to southeast.

21. Of all the earth's water, only 3 percent is fresh.

22. The full quote is: "The map is not the territory." We suppose this is why sitting at home looking at maps is not as much fun as actually going places.

23. Brazil has the highest murder rate in the world. For every 100,000 people, 104 were murdered in 1983, the last year for which figures are available. The lowest rate in the world was in the Maldives, where there has not been a murder since these islands became independent in 1965. The Bushmen of South Africa are so little inclined to murder that when one of them committed the crime in the 1930s, he voluntarily banished himself from human company and has not been heard from since.

And the United States? Our rate was 10 murders per 100,000 people. Eighty percent of the countries that report murder rates have lower rates than ours. Neighboring Canada, for example, whose history and culture are in many ways similar, has a rate of only 2 per 100,000.

24. Probably the two most important cities associated with cherry blossoms are Tokyo and Washington, D.C., both of which put on impressive floral displays every year. What connects the two is David Fairchild, the naturalist, plant explorer, and professional introducer of exotic species, who brought the cherry trees from Tokyo, where they had impressed him on his travels around the world in his own Chinese junk, to the U.S. capital. Fairchild is one of hundreds of people throughout history, many of them forgotten, who changed the look of one place by introducing something from another. He lived in Coconut Grove, Florida, and even though he was married to a daughter of Alexander Graham Bell, inventor of the telephone, he refused to have a free phone installed because he considered that instrument an invader of privacy. But we digress.

25. Every map should have a title (such as "Africa," or "Distribution of Lung Cancer in North America by Counties") telling what is represented; a scale, showing the ratio of map distances to real-world distances ("One centimeter equals 100 kilometers," or a bar graph); a legend, to explain the symbols employed on the map (dot with a star inside = a capital city, tree = forest); latitude and longitude, or other method of orientation (on maps of small areas, a "north" arrow serves); indication of type of projection (since the world is round and maps are flat, any map distorts, and the cartographer should tell which kind of distortion the map reader will be dealing with).

26. Babson College has the world's largest relief map of the United States. Thirty by sixty feet, it is viewed from a surrounding balcony and shows this country in perfect, precise detail. It is worth a trip.

27. The surplus heat from the central zone flows toward the poles via atmospheric and oceanic currents. The dividing line averages out to about thirty-eight degrees north latitude, the region of Madrid, Philadelphia, and Beijing.

28. The Gateway National Seashore at Jamaica Bay, ten miles from Times Square as the crow flies, has a subway traversing it. This area was about to become a city dump in the late 1950s, but was cleaned up and planted with beach grass. It has become a wild, natural seaside park where 310 species of birds were sighted in 1984.

29. The whole earth's albedo is about thirty-three; that is, some 33 percent of incoming solar energy is reflected back into space. The main reflectors are the bright white cloud tops you see from airplanes.

30. The actual surface of the earth, as astronomers define it, is the outer edge of the atmosphere, about one hundred miles above our heads. The atmosphere is an integral part of the earth. We and all our works are best thought of as boundary-layer phenomena— existing at the *bottom* of the atmosphere on top of the solid earth, and *alongside* the liquid earth, that is, the oceans. The layer we inhabit, up to about twelve miles over our heads, is the troposphere, and unless you have flown on a supersonic aircraft or ascended in some high-technology balloon, you've never left it. The layer above is the stratosphere.

31. The carbon in your left ring finger, and all the other fingers, came from what you've eaten over the last few days, weeks, and months. Before that, the carbon was something else, and before that, something else again. The world's basic stock of elements has been recycling since the earth began. Some of the carbon in your finger might have been in a tree five thousand years ago, and before that, some of you may have been in a dinosaur, or a hunk of limestone, or a wisp of natural gas. You've always been around, you know—in one form or another.

32. The Arab maps of old were beautiful, with elevations noted in various colors—purple representing the highest—and cities set out with dots of real gold leaf. They were also quite accurate. But Western explorers and navigators, not often reading Arabic, used them upside-down. Arab cartographers always placed south at the top. Why? See the answer to question 35.

33. The most densely populated continent is Europe, with 173 persons per square mile; the longest river system, the Nile, 4,187 miles long, is in Africa; the lowest inland point you can drive a car to is the Dead Sea, 1,302 feet below sea level, in Asia; the highest waterfall is Angel Falls in South America; the largest island is Greenland near North America; and the South Pole is in Antarctica.

34. The only point from which you can see both Atlantic and Pacific oceans is on the crest of the volcano Irazú, in Costa Rica, Central America. It is more than 11,000 feet above sea level and has become something of a tourist attraction, which is why many tourists were killed when *Volcan Irazú* erupted unexpectedly in the early 1960s.

35. Don't ask us.

36. Sudan
Brazil
USSR
USSR
Canada

37. Nigeria
Brazil
USSR

38. The largest pyramid is that of Quetzalcoatl, in Mexico, whose volume is 3.3 million cubic meters. The Pyramid of Cheops has only 2.5 million.

39. Goshen is the northeastern part of the Nile Delta, in Egypt.

40. Colombia, in South America, which has guerrilla violence, dope-smuggling violence, a history of political violence, and a lot of just plain nasty crooks, seems to be the most violent nation, according to a 1985 comparison of acts of mayhem.

41. The number of free, sovereign countries is hardly carved in stone. In 1950, for example, only thirty-six years ago, Africa had only four independent countries: Ethiopia, Liberia, the Union of South Africa, and Egypt. All the rest of that continent was European colonial territory. Today things are different, and it stacks up this way:

1–f	5–b
2–a	6–d
3–e	7–c
4–g	

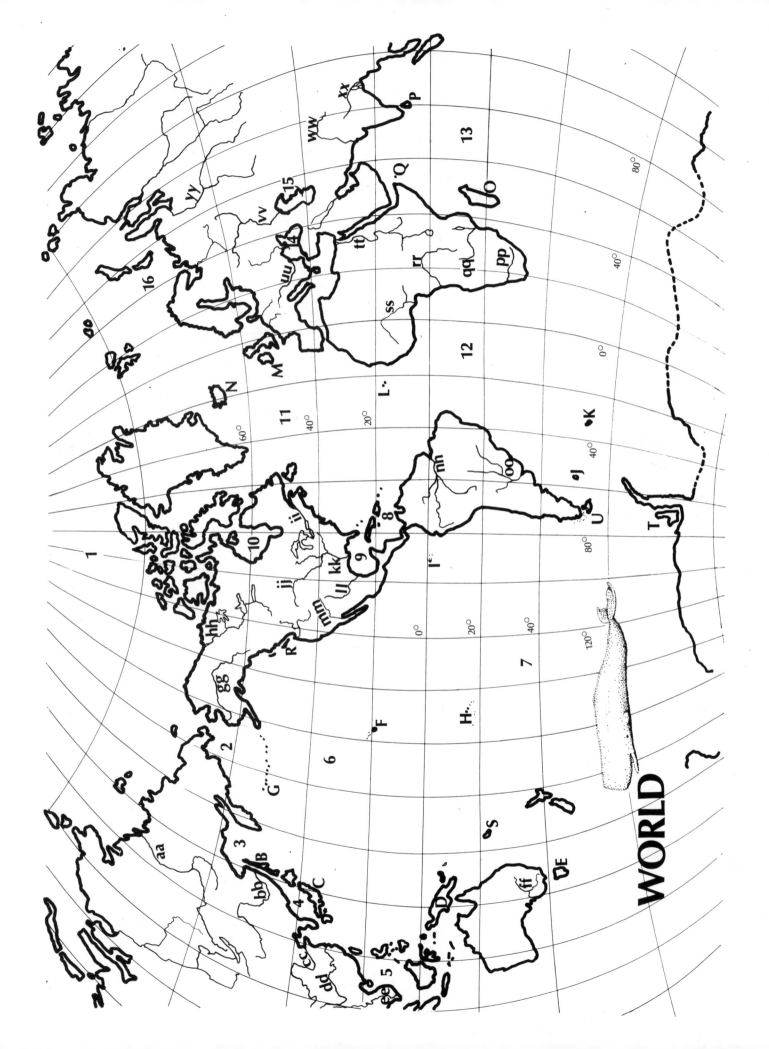

WORLD

THE WHOLE EARTH

Answer the following questions. The accompanying world map may help.

COUNTRIES AND CONTINENTS

1. Excepting Canada and Mexico, what country is closest to the United States?

2. What three South American countries are along the equator?

3. What two North American countries are along the Tropic of Cancer?

4. What four South American countries are along the Tropic of Capricorn?

5. What three European countries are along the Prime Meridian?

6. What five African countries are along the Prime Meridian?

7. What six African countries are along the equator?

8. What are the three countries along 180 degrees longitude?

9. Name the countries found at the following latitude-longitude designations:

a lat. 40°N long. 100°W
b lat. 20°N long. 100°W
c lat. 40°N long. 100°E
d lat. 20°N long. 100°E
e lat. 30°S long. 60°W
f lat. 60°N long. 30°E
g lat. 30°N long. 60°E
h lat. 35°N long. 140°E
i lat. 35°S long. 140°E
j lat. 50°N long. 10°E
k lat. 52°N long. 10°W
l lat. 20°N long. 50°E

SEAS, ISLANDS, AND RIVERS

Locate the following on the world map (seas are marked with numbers, islands with letters, rivers with double letters):

1 South Pacific Ocean __7__
2 North Atlantic Ocean __11__
3 Gulf of Mexico __9__
4 New Guinea __D__
5 Hawaii __F__
6 Mississippi River __Kk__
7 Amazon River __nn__
8 Nile River __tt__
9 Rio Grande __ll__
10 Madagascar _____
11 Galápagos Islands __I__

12 Black Sea _14_
13 Sri Lanka _P_
14 Missouri River _jj_
15 Tahiti _A_
16 Tasmania _E_
17 Ireland _M_
18 Honshu _C_
19 South China Sea _5_
20 Iceland _N_
21 Yukon River _hh_
22 Amur River _____
23 Mekong River _ee_
24 Bering Sea _2_
25 Falkland Islands _J_
26 Hudson Bay _10_

27 Orange River _____
28 Niger River _qq_
29 Indus River _xx_
30 Colorado River _mm_
31 Socotra _____
32 Shemya _____
33 South Georgia _____
34 Cape Verde Islands _____
35 Sea of Okhotsk _3_
36 Murray-Darling River _____
37 Ganges River _ww_
38 Mackenzie River _gg_
39 New Caledonia _SJJ_
40 Tierra del Fuego _U_

THE WHOLE EARTH MAP QUIZ

COUNTRIES AND CONTINENTS:

1. The USSR. In the Bering Strait, Soviet- and United States–owned islands are only three miles apart.

2. Ecuador, Colombia, Brazil.

3. Mexico and the Bahamas.

4. Chile, Argentina, Paraguay, Brazil.

5. United Kingdom, France, Spain.

6. Algeria, Mali, Burkina Faso, Togo, Ghana.

7. Gabon, Congo, Zaire, Uganda, Kenya, Somalia.

8. United States, USSR, Fiji.

9.
  a United States
  b Mexico
  c China
  d Thailand ("Golden Triangle")
  e Argentina
  f USSR (Leningrad)
  g Iran
  h Japan
  i Australia
  j West Germany
  k Ireland
  l Saudi Arabia

SEAS, ISLANDS, AND RIVERS:

1–7	21–gg
2–11	22–bb
3–9	23–ee
4–D	24–2
5–F	25–J
6–kk	26–10
7–nn	27–pp
8–tt	28–ss
9–ll	29–ww
10–O	30–mm
11–I	31–Q
12–14	32–G
13–P	33–K
14–jj	34–L
15–H	35–3
16–E	36–ff
17–M	37–xx
18–C	38–hh
19–5	39–S
20–N	40–U

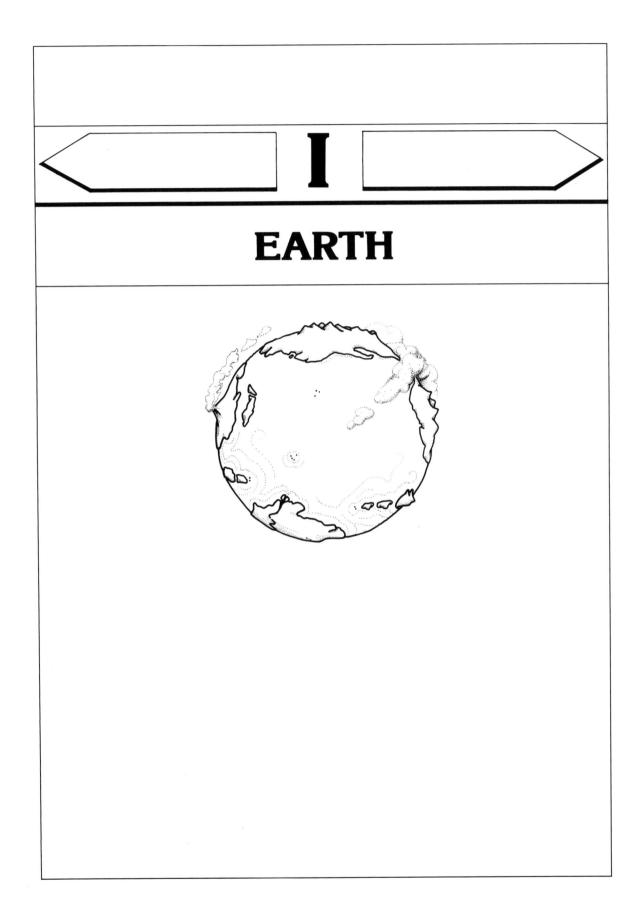

I
EARTH

LANDFORMS

In the blackness of space, Earth is the blue planet. Its oceans and thick atmosphere of breathable gas make it unlike any other in our solar system. Mars, Venus, Mercury, and our moon also have solid crusts and were shaped by similar forces: folding, faulting, and volcanism. But they have no oceans, no rivers, no rain. Their surfaces are dry and pocked with impact craters. Venus has acid fogs.

The Earth's unique landforms are the product of its internal dynamics combined with the movements of its special fluids: volcanoes, the drifting of continental plates, the rain and storms, snow and runoff. The land is various. Some is lofty and rough, some is low and spread flat. Water has cut grooves in it and still flows through many of them. Mountains build up along the boundaries of tectonic plates or stand where volcanoes once were active. Then they are reduced by erosion from rain and runoff. Left to itself, land, like water, will seek its own level. It just takes longer.

Scenery—which we pay large sums to visit or live near—is just another name for attractive landforms.

CONTINENTS

1. Which great landforming events does the Bible describe?

2. One of the revolutionary ideas in the history of scientific thought is the notion of plate tectonics, or continental drift. What does this theory claim?

3. What are the biggest landforms?

4. In the South Pacific, there are three types of island. They are very different, each determined by the sort of rock it is made of. One type is the "little continent"—islands made of the same stuff as all the continents. Think of the Pacific islands you've seen or read about, and name the other two kinds of island found there.

5. Which continent has the fewest hills and the lowest average elevation?

6. On what continent is the farthest spot below sea level?

MOUNTAINS

1. Why are the New England Appalachians craggy, whereas the southern Appalachians are more rounded?

2. After a volcanic eruption, the land thereabouts turns sterile black, a result of blast, ash, and lava. Do the plants and animals ever return?

3. Where you find mountains, you find valleys in between.
   a What valley produced Valley Girls?
   b Where was the Valley of the Dolls?
   c Who lived in the Big Valley?
   d Where did Silicon Valley get its name?
   e Where is Moon Valley?

4. Near which ocean will you find the most volcanoes?

5. The mid-ocean ridges extend across the floors of every ocean. They connect, forming a ridge forty thousand miles long, and rise as high as twenty thousand feet from the ocean floor. Why are these greatest of mountain ranges on earth under the oceans?

6. Name the highest mountain in each of these places: Antarctica, the United States, Canada, Mexico, Colorado, South America, Africa, Australia, New Zealand, Turkey.

7. Name the highest alp.

8. What caused the loudest noise ever heard on earth?

9. Match these countries with their mountains:
1	England	a	Urals
2	France	b	Grampians
3	Italy	c	Pennines
4	Romania	d	Pindus Range
5	USSR	e	Alps
6	Switzerland	f	Carpathians
7	Bulgaria	g	Apennines
8	Greece	h	Massif Central
9	Norway	i	Rhodopes
10	Scotland	j	Kjølen Mountains

10. What are the Arlberg, the Brenner, the Saint Bernard, and the Simplon?

11. Where are the Mountains of the Moon?

12. Here is a wonderful phrase that might apply to your last party: "mass wasting." What does it really refer to?

RIVERS

1. Complete this quotation from Herodotus: "Egypt is the gift of the _____."

2. What rivers do is discharge water from continents to large bodies of water. If we asked what U.S. river discharges the most water, we assume you'd answer, "The Mississippi." So we won't. Instead, which U.S. river discharges the second most?

3. About 212,000 cubic feet of water pours over Niagara Falls every second. Guess where this ranks Niagara in volume of discharge:
First in the world
Fourth in the world
Twelfth in the world
Twenty-second in the world

4. Name the great cities you'd find on these rivers:
Hudson	Manzanares
Thames	Hooghly
Seine	Yamuna
Charles	Han-gang
Almendares	Parramatta
Guaire	Tiber

5. What is the longest river in Canada?

6. How was the Mississippi Delta made?

7. Rivers don't always go where you want them to, which is why humans invented canals. Where is the longest canal?

8. What is the great river of Syria that you heard so much about in school and is so famous it got a cracker named for it, but has not been heard from since?

9. Pittsburgh, Cincinnati, Evansville, and Cairo—what do they have in common?

10. What is a "lost river"?

11. For generations, explorers sought the source of the Nile, and found it at last. But where is the source of the Mississippi?

12. What makes a waterfall a waterfall?

13. How did Angel Falls, the highest in the world, get its name?

14. Which landform is known as "The Strong Brown God"?

15. Many places get their names from the names of prominent nearby landforms. India is one; what is the source of its name?

16. Match the river with the dam:
    1 Tennessee   a Hoover
    2 Columbia   b Shasta
    3 Missouri   c Pickwick
    4 Colorado   d Fort Peck
    5 Sacramento   e Grand Coulee

17. Think of all the galleries and bookshops and cafes along the Boulevard Saint Michel on the Rive Gauche, or Left Bank. Is the "left bank" of the Seine River the north or south side of the channel?

18. Match these rivers with the bodies of water they flow into:
    1 Mississippi   a Baltic Sea
    2 Ohio   b Gulf of California
    3 Columbia   c Mediterranean Sea
    4 Fraser   d Dead Sea
    5 Potomac   e Arabian (Persian)
    6 Zaire   Gulf
    7 Rhine   f Pacific Ocean
    8 Vistula   g Gulf of Mexico
    9 Saskatchewan   h Atlantic Ocean
    10 Huang He   i Hudson Bay
    11 Amazon   j Bay of Bengal
    12 Colorado   k Mississippi River
    13 Indus
    14 Tigris
    15 Jordan
    16 Nile

19. What factor determines whether running water is able to carry boulders, or gravel, or only sand?

20. Why do rivers flood?

21. We sometimes hear politicians champion a pork-barrel dam project on the grounds that it will provide recreation, hydroelectric power, irrigation water, *and* flood control. Why is this bull?

22. Where can you find fresh water two hundred miles out at sea?

LAKES

1. Where are the women strong, the men good-looking, and all the children above average?

2. What happens eventually to every lake?

3. Which is the biggest lake in the world?

4. Which is the world's largest artificial lake, by surface area?

5. Which of the North American Great Lakes is the largest?

6. Which is the world's deepest lake?

7. Why are there more lakes in the northern United States than in the south?

8. What do Lake Eyre, the Dead Sea, the Salton Sea, and the Caspian Sea all have in common—besides water?

9. What do the Chicago Bears, Cleveland Browns, and Toronto Argonauts have in common?

10. Where are the Bitter Lakes?

11. The Germans call it the Bodensee. What do Americans call it?

SEASHORES

1. Beach sand is the end product of nature's forces pounding on rocks and shells, constantly reducing their size. What happens to beach sand, which presumably is exposed to the same forces?

2. Match the kind of beach material with the coast you'd probably find it on:
    1 Quartz sand   a England
    2 Coral sands   b Bermuda
    3 Shingle   c Florida Panhandle
    4 Lava sands   d Southeastern Florida
    5 Shell   e Island of Hawaii

3. What is the main difference between the summer beach and winter beach around Cape Cod?

4. How is the sand on most beaches like a river?

5. Upon what body of water does San Francisco's North Beach front?

6. How many of the eleven beaches named in the Beach Boys' song "Surfin' USA" can you remember? Hint: They're all in one state.

7. Only one U.S. state has a town named Beach. In what state is Beach?

LANDFORMS

CONTINENTS

1. Only two landforming events are found in the Bible: the Creation, in which all is made in six days, and the Flood, in which the earth is deluged by forty days of rain and almost everybody drowns. It has been left to the earth sciences to explain everything else. More than any other scientists in the past one hundred years, the earth scientists have challenged the fundamentalists' ideas.

2. Plate tectonics suggests that the crust of the earth is divided into separate sections, or plates. The plates are rigid, but along their edges they jostle, drift apart, or crunch together, in the same way that the cracked shell of a hard-boiled egg can shift and grate. Heat from inside the earth wells up along some of the cracks, melting the crust and feeding magma—melted rock—to volcanoes. In other places, the plate edges scrape past each other or collide, and one dives beneath the other. Those are the three kinds of plate boundaries: (a) where plates spread and lava fills the void; (b) where plates collide and one dives under; and (c) where plates slide past each other.

   Continents and plates are not the same thing; continents seem almost to be accidental riders on the plates. Plates can spread apart beneath a continent, as seems to be happening in East Africa, where the rift valleys are the breaks; carry a continent to a collision with another, as seems to have been happening during the last 10 million years, as India crunches against Asia, pushing up the Himalayas; or crack continents

where they slide past each other, as at the San Andreas Fault in California.

3. The continents and ocean basins are the biggest landforms. Beneath their surface layers of sediment, the continents are mainly composed of granites, crystalline rocks of gray, white, or pink, which also make good bank facades and tombstones. The continents' average elevation is 2,870 feet. They stand high because their rock is lighter (of lower density) than that of the ocean basins. These are made of basalts, black, fine-grained rocks that bubbled up originally as magma—melted rock—from such undersea volcanoes as formed Hawaii. The average depth of the oceans is 12,430 feet, making them much deeper than the continents are high. The amount of water with which the earth has been supplied is just enough to fill up the ocean basins to the edge of the continents, with just a little left over to make waves. The continental shelves are usually covered by 200 to 300 feet of ocean.

4. The other two South Pacific island types are (a) volcanic, great piles of basalt rising from the ocean floor, usually called "high islands," and including the Hawaiians, the Galápagos, and the Marquesas, and (b) coral islands, made of limestone created as reefs and atolls by living coral polyps. The greatest cluster of coral islands is the Tuamotu group.

5. Australia, a place of low, worn mountains and vast sandy plains, is the lowest continent.

6. The spot farthest below sea level is a depression 8,100 feet below sea level in Marie Byrd Land in Antarctica. But it is filled with ice. The bottom of Lake Baikal, in Siberia, is the second deepest, at minus 4,277 feet.

MOUNTAINS

1. The southern mountains were untouched by the ice sheets that eroded the New England mountains. In the south, the rock has remained covered by topsoil, and a humid climate encourages the growth of trees. You can tell by their low, rounded, worn-down shape that the southern mountains are very old.

2. Plants and animals begin to recolonize almost immediately following a volcanic eruption. When things cool down, animals amble in. Some dig through the ash, bringing up richer soil from beneath. Windblown seeds arrive and take root. At Mount Saint Helens, every species that occupied the region before the 1980 eruption was cheerfully living there five years later. But it will take centuries before the soil is rich enough once more to support an authentic-looking Pacific Northwest forest. Mount Saint Helens is still active, and geologists expect it to rebuild slowly over the next few hundred thousand years.

3. **a** Valley Girls came from the San Fernando Valley.
   b We don't know where the Valley of the Dolls is, and if you do, you should be ashamed of yourself.
   c The Big Valley was inhabited by Barbara Stanwyck, in the 1960s TV show of the same name. It was a sort of downslope *Bonanza*.
   d Silicon Valley, in California, the string of office and industrial parks extending from near Palo Alto to beyond San Jose, is named for the stuff semiconductors are made from.
   e Moon Valley is in Rabun County, Georgia, where *Deliverance* was filmed and the Foxfire books originated.

4. The rim of the Pacific Ocean is called the Ring of Fire for its many volcanoes. They extend from Antarctica through the Andes, along North America and then across the Aleutians to Japan, the Philippines, and New Zealand.

5. Since erosion here is very slow, most still carry the shape of their birth. They are jagged volcanic mountains.

6. Antarctica, Vinson Massif, 16,864 feet; the United States, Mount McKinley, 20,320 feet; Canada, Mount Logan, 19,850 feet; Mexico, Citlaltépetl, 18,701 feet; Colorado, Mount Elbert, 14,433 feet; South America, Aconcagua, in Argentina, the highest in the hemisphere at 22,834 feet; Africa, Mount Kilimanjaro, in Tanzania, 19,340 feet; Australia, Mount Kosciusko, 7,316 feet; New Zealand, Mount Cook at 12,349 feet; Turkey, Mount Ararat, 16,945 feet.

7. The highest alp is Mont Blanc, in France and Italy, at 15,771 feet; it has a very chic fountain pen named for it. The Matterhorn in Switzerland and Italy is second at 14,690 feet.

8. No one can be certain about the loudest sound ever heard. But the loudest ever actually recorded came from a tiny island called Krakatoa, between Java and Sumatra, one day in August 1883. Krakatoa turned out to be a volcano, and began throwing ash, pumice, and blocks of lava from its slowly emptying crater. Finally, the crater collapsed, and tons of ocean water poured onto the superheated lava, like dishwater into a big hot skillet—followed by about two-thirds of the island itself. The noise was heard so clearly 5,000 kilometers away that an army commander thought it was gunfire and deployed his troops. The explosion caused a tidal wave forty feet tall and a stupendous atmospheric shock wave so powerful it circled the earth seven times.

9.
1–c	6–e
2–h	7–i
3–g	8–d
4–f	9–j
5–a	10–b

10. They are passes through the Alps.

11. The Mountains of the Moon are on Earth. This is the name Ptolemy gave to the legendary mountains of Central Africa, and sure enough, they are found on the Uganda-Zaire border. But today they are called the Ruwenzori. Their highest peak is Mount Stanley (16,763 feet), named for Henry Morton Stanley, the British-American newspaper writer and explorer who, in 1889, was the first European to reach them.

12. Mass wasting is what happens on hillsides; it is the downslope erosion caused by gravity in the form of rockslides, mudflows, and soil creep. "Soil creep" is another nice phrase, but that's another story. Gravity never sleeps.

RIVERS

1. "Nile."

2. The Columbia, in Oregon and Washington, with 262,000 gallons per minute on the average.

3. Niagara is fourth in the world. First is the former Stanley Falls, now called Boyoma Falls, on the former Congo River (now Zaire River), with 600,000 cubic feet per second.

4. New York, London, Paris, Boston, Havana, Caracas, Madrid, Calcutta, Delhi, Seoul, Sydney, Rome.

5. The Mackenzie, in far northwest Canada, at 2,635 miles.

6. It is still being made. Every day, tons of mud come down the river from the basins of the Mississippi, Ohio and Missouri rivers and settle to the bottom as the power of the river peters out. The delta is ever extending itself farther into the Gulf of Mexico.

7. The longest single canal in the world is the 141-mile Baltic–White Sea canal, in the USSR. The USSR also boasts the longest canal system. On it, using the rivers it interconnects, you can go from Astrakhan on the Caspian Sea, up the Volga and all the way to Leningrad, about 1,850 miles.

8. The Euphrates is the source of much of our civilization. Today, dams on the Euphrates River are a source of conflict between Syria and Iraq. The Euphrates cracker, as we recall, is sprinkled with sesame seeds.

9. The Ohio River. The Cairo in Illinois is the one we mean.

10. A lost river vanishes into the earth partway along its course. This happens most often when a river plunges into a limestone cavern system, only to reappear somewhere else.

11. The Mississippi is born as an insignificant little brook at Lake Itasca, Minnesota, more than 2,500 miles north of its mouth in Louisiana. It was "discovered" in 1832 by an explorer named Henry Schoolcraft, accompanied by an Indian guide who already knew where it was. The lake's name comes from the middle syllables of *veritas caput*—Latin for "true source." Rivers begin in various ways: as outlets from little lakes, like this; or as springs; or as water-worn grooves in rock where rain collects and begins to flow.

12. When a stream runs over a hard bedrock and then strikes a spot where the bedrock is softer, it will wear down the softer rock faster, making it lower and creating a drop, or fall. Niagara Falls, for example, is where the Niagara River falls over the cliff created by tough dolomitic rock of the Niagara Escarpment, which is visible in one form or another all the way to Wisconsin. Most of the world's falls can be explained thus.

13. In 1935 an eager American bush pilot and gold prospector flew up the canyons of the Guiana Highlands in Venezuela and discovered more than a dozen of the world's highest falls. His name was Jimmy Angel.

14. The Niger River, in West Africa, which flows from the mountains of Guinea inland toward the Sahara and then turns south to join the sea in eastern Nigeria, is known as "The Strong Brown God." Take a point if you knew the god was a river. The Niger connects and divides six countries.

15. India's name comes from the great Indus River, which issues from a single tiny spring in Tibet called *singi kabab*, or "the mouth of the Lion." But life is strange: The Indus flows through India-controlled Kashmir before spending its career in Pakistan. The waters of the Indus do not bathe the fertile plains of India; instead they are the lifeblood of India's bitterest enemy.

16. 1–c
    2–e
    3–d
    4–a
    5–b

17. The Left Bank is on the south; right and left banks are determined by facing downstream.

18. 1–g   9–i
    2–k   10–f
    3–f   11–h
    4–f   12–b
    5–h   13–j
    6–h   14–e
    7–h   15–d
    8–a   16–c

19. The speed of the moving water: One foot per second will move gravel, four feet per second will turn a two-pound rock, and twenty-five feet per second will roll boulders.

20. Rivers flood because more water gets into them than they are used to handling; the usual way is via extra-heavy rains or rapid snowmelt. A burst dam can do it, too.

21. The politician cannot be telling the truth because his claims are mutually exclusive. Think about managing the water level. For recreation, you need to keep the water level constant, for boat ramps and beaches. For hydroelectric power, you must let water out through the dam when you need electricity, which drops the level. For irrigation water, you must allow water through when crops require it. And for flood control, you need to keep the dam empty. The truth is, dam projects usually benefit only a few users, and the question for the voter is, "Whose goals do I sympathize with?"

22. There is fresh water two hundred miles out at sea from the mouth of the Amazon River, in South America. The Amazon, amazingly, carries one-fifth of all the fresh water that runs off the surface of the earth—more than the Nile, Mississippi, and Yangtse combined. The Amazon's fresh water is lighter than the salty Atlantic Ocean, so the Amazon's water glides out over the ocean surface, mixing only gradually.

LAKES

1. Lake Wobegon, Minnesota, on the radio show *Prairie Home Companion*.
2. Every lake eventually fills with sediment, shrinks, and becomes dry land.
3. The Caspian Sea, at 143,550 square miles, is the world's biggest, if you define "lake" as a body of water entirely surrounded by land. The Caspian, which is salty, and shrinking because it loses more water through evaporation than the Kura, Terek, Ural, and Volga rivers bring in, is bounded on three sides by the USSR and on the fourth by Iran.
4. Lake Volta, in Ghana, covers 3,275 square miles and was formed by the completion in 1965 of the Akosombo Dam.
5. Lake Superior. It is also the second biggest in the world, at 31,800 square miles.
6. The world's deepest is Lake Baikal or Baykal in Siberia. It is 5,315 feet deep.
7. The north is where the Ice Age glaciers cut lake beds. Minnesota, for instance, has more than ten thousand lakes. There are 2 million plus in Quebec.

   But there are lakes in the South, including one of the biggest, Lake Okeechobee, in Florida. Lake Okeechobee is unusual in that it is not drained by a proper river with a proper riverbed, but by the Everglades or "River of Grass," as the author Marjorie Stoneman Douglas called it, a slowly moving sheet of water encircled by a slightly higher limestone ridge on which are built Palm Beach, Fort Lauderdale, and Miami.
8. Their surfaces are all below sea level.
9. All three teams play in stadiums along the shore of a Great Lake—Michigan, Erie, and Ontario, respectively.
10. The Bitter Lakes are depressions of salty water that were connected by excavation of the Suez Canal in 1869.
11. Americans take their lead from the French (in this case) and call it Lake Constance.

SEASHORES

1. Waves carry the lighter grains away and toss heavier ones back onto the high beach. But once a grain of sand is established on a beach, it is destined to remain a grain of sand almost forever because further wear is extremely slow. "A grain of sand is virtually immortal," writes naturalist Peter Farb. "Each grain at the beach is cushioned against further abrasion by being surrounded by a film of water, whose volume is approximately equal to the volume of the grain. Along the lower beach the grains appear to be closely packed, but in fact they never touch."
2. 1–c
   2–b
   3–a (small, flat stones)
   4–e (volcanic)
   5–d
3. During the summer on Cape Cod, gentle waves deliver tons of sand landward, building up the beach. Storms arrive in winter, and waves pound and claw away the summer's beach. Each year, the beach is built up anew.
4. The sand on both coasts of the mainland United States is flowing southward along the beach, much like a river with the ocean as one bank and land as the other. It is carried by the "longshore current."
5. North Beach, the historically Beat and Bohemian neighborhood of San Francisco, is in the middle of the city, fronting upon no body of water at all.
6. The beaches in the Beach Boys' tune were La Jolla, Doheny, Redondo Beach, Los Angeles (City Beach), Corona Del Mar, Santa Cruz, Manhattan Beach, Pacific Palisades, San Onofre, Sunset Beach, Ventura County Line Beach—the latter now called Leo Carillo Beach, after the actor who played the Cisco Kid's pal, Pancho.
7. America's only Beach is in landlocked western North Dakota.

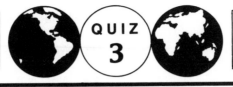

OCEANS

Oceans occupy about 70 percent of the earth's surface. Strangely enough, the human body is also about 70 percent water. The fact is probably not significant, but that does not lessen its interest as coincidence and as an occasion to ponder our place in the world and our relation to these most mysterious volumes of geography that lie all around us.

1. On the average, about how deep is the ocean?
   a A quarter mile deep
   b Two and a quarter miles deep
   c About seven miles deep

2. Which is larger—Asia or the Pacific Ocean?

3. What makes waves?

4. Where does the Flying Dutchman sail?

5. Why do sea levels change? (There are many reasons.)

6. Where is the oceans' deepest spot?

7. What is the name of the Florida island Richard Nixon vacationed on when he was president?

8. How many islands are there in the well-known salad dressing, and where are the originals?

9. The ocean amounts to about (choose one: *55 percent, 75 percent, 95 percent*) of the life-supporting space on earth.

10. If you combined all the world's rivers and then multiplied the volume of their flow by five, you'd have the flow volume of what?

11. The Gulf Stream flows essentially (*north, south, east,* or *west*?) along the U.S. coast.

12. Built for the CIA by Howard Hughes, for an attempt at salvaging a Soviet submarine from the floor of the Pacific ocean—what is it?

13. Where is the world's largest tropical coral-reef complex?

14. Match the coast with its location:
   1 Ivory Coast   **a** U.S. Gulf of Mexico
   2 Barbary Coast   **b** Morocco, Algeria,
   3 Skeleton Coast   Libya
   4 Tanga Coast   **c** West Africa
   5 Gold Coast   **d** U.S. East Coast
   6 Redneck Riviera   **e** France, Italy
   7 Riviera   **f** Namibia
   8 Mosquito Coast   **g** East Africa
                          h Central America

15. Where are:
  a The Georges Bank
  b The Grand Banks
  c The Bank of England

16. Match these straits with what they separate:

1 Cook Strait		**a**	Cuba and Haiti
2 Bass Strait		**b**	Luzon and Taiwan
3 Strait of Malacca		**c**	Ireland and Wales
4 Korea Strait		**d**	Lake Huron and Lake Michigan, or lower Michigan and upper Michigan
5 Sōya Strait			
6 Luzon Strait			
7 Strait of Hormuz			
8 Palk Strait		**e**	Vancouver Island and the Olympic Peninsula
9 Dardanelles, Bosporus			
10 Strait of Messina		**f**	Denmark and Sweden
11 Strait of Bonifacio			
12 Strait of Gibraltar		**g**	Denmark and Norway
13 Strait of Dover		**h**	France and England
14 Skagerrak		**i**	Spain and Morocco
15 Kattegat		**j**	Sardinia and Corsica
16 Saint George Channel		**k**	Calabria and Sicily
17 Windward Passage		**l**	Black Sea and Mediterranean
18 Juan de Fuca Strait		**m**	Sri Lanka and India
19 Straits of Mackinac		**n**	Iran and Oman
		o	Sakhalin and Hokkaidō
		p	Sumatra and Singapore
		q	Tasmania and Victoria, Australia
		r	North and South Island, New Zealand
		s	Korea and Japan

17. What do the California Current, Peru Current, Benguela Current, and Canary Current have in common?

18. Name an island that was a prison colony for a prison colony.

19. What was the *Bounty,* Captain Bligh's and Fletcher Christian's doomed ship, carrying and to where?

20. What people discovered the biggest island in the Atlantic Ocean?

21. How do you tell a big island from a continent?

22. What is the largest island in Canada?

23. What is the largest island in the Pacific Ocean?

24. What is the world's second largest island?

25. How many different ways can islands form?

26. Is there really such a thing as a desert island?

27. Would you want to live on a terminal moraine?

28. Do island nations become sea powers?

29. Where is the deepest place in the Atlantic Ocean?

30. Who introduced the ukulele to Hawaii?

31. What is the Archipiélago de Colón, an offshore part of Ecuador, better known as?

32. Name the country that claims to own the following islands:
  a Easter Island (Rapa Nui)
  b Pitcairn Island
  c Prince Edward Islands (South Indian Ocean)
  d Macquarie Island
  e Jan Mayen Island
  f South Georgia
  g Réunion
  h Guadalcanal
  i Yap
  j Grand Cayman
  k Bermuda
  l Socotra
  m Saint Helena
  n Montecristo
  o Staten Island

33. Which, according to James Michener, is "the most beautiful island in the world"?

34. Which, according to Christopher Columbus, is "the most beautiful island ever seen"?

35. These are two places named by the Japanese— Iwo Jima and Hiroshima. How do we know they are islands?

36. What island is in the Sargasso Sea?

37. What is islomania?

38. Match these islands with the group they are members of:

1	Abaco, New Providence	a	Netherlands Antilles
2	Oahu, Maui	b	Balearic Islands
3	Makatéa, Bora-Bora	c	Aleutian Islands
4	Truk, Yap	d	Channel Islands
5	Jersey, Guernsey	e	Caroline Islands
6	Majorca, Minorca	f	Society Islands
7	Aruba, Bonaire	g	Hawaiian Islands
8	Kiska, Unimak	h	Bahamas

39. What did Leicester Hemingway, Ernest's younger brother, have to do with a certain island republic?

40. What islands are Pinta, Pinzon, Santa Cruz, and Isabela?

41. Match these well-known bays with their countries:

1	Bantry Bay	a	Australia
2	Table Bay	b	China
3	Hudson Bay	c	West Germany
4	Guanabara Bay	d	Japan
5	Bo Hai	e	South Africa
6	Ise Wan	f	Brazil
7	Kiel Bay	g	Canada
8	Botany Bay	h	Ireland

42. Match the island groups with their nearest continent:

1	Galápagos	a	Asia
2	Balleny	b	Australia
3	Bahamas	c	Europe
4	Kurils	d	North America
5	Furneaux	e	Antarctica
6	Orkneys	f	South America

43. On the ocean floors are vast regions so uniformly flat that in one thousand square miles they may have less than ten feet of relief. They are greater in fact than those wide fields of wheat you see in the U.S. Midwest. These impressive undersea "prairies" are called _____.

44. In Allentown, Pennsylvania, is the world's largest indoor _____ machine. Fill in the blank.

45. Tuna is "_____ of the sea"?

OCEANS

1. The oceans average about two and a quarter miles deep.

2. The immense continent of Asia, at 17,085,000 square miles, is only a third the size of the Pacific Ocean.

3. Waves, except for those minor ones caused by freighter wakes, surfacing whales, and floundering swimmers, are caused by wind. Storms cause the big waves of the open ocean, which can travel great distances. A storm near Iceland, for instance, can make breakers on the northern coast of Brazil. Waves from Antarctic storms have damaged ships anchored in Los Angeles harbor.

4. The Flying Dutchman is said to sail empty and forever off the Cape of Good Hope, beyond Africa's southern coast.

5. Here are five of the reasons:

   a Over the long term, the volume of water in an ocean basin can be changed by eruptions of volcanoes along the mid-ocean ridge system.

   b Ocean water becomes glacier ice during an ice age, thus making sea levels drop; during the last ice age, they were two to three hundred feet lower than they are now.

   c When the land sinks, the oceans seem to rise. When humans began pumping oil from Long Beach, California, harbor, the land was no longer supported by the oil deposit and began to sag. The sea thus seemed to rise. The sinking was stopped by pumping water in to replace the oil.

   d Weather. High atmospheric pressure lowers the ocean surface, simply by pressing down on it. Low pressure allows sea level to rise. Hurricanes have

such low pressure that a bulge of water up to forty feet high travels along beneath them.

   e Tides, of course.

6. There is a trench, the Mariana Trench, in the Pacific Ocean off the Mariana Islands, that dives to 36,201 feet below the sea's surface—via a slope that seldom exceeds two degrees. It is the deepest known spot.

7. Key Biscayne. Bebe Rebozo lived there too.

8. Thousand Island dressing was named for the Thousand Islands of the Saint Lawrence River, just below Lake Ontario. There's a place on the lower Gulf coast of Florida called the Ten Thousand Islands, but dressing named for it was found too thick to pour.

9. The ocean has at least 95 percent of the planet's life-supporting space, according to some experts.

10. The Gulf Stream.

11. The Gulf Stream flows essentially northward along the coast of the U.S. before turning eastward toward Europe.

12. The *Glomar Explorer*, a drillship now used to explore the oceanbeds.

13. The greatest reef of all is Australia's Great Barrier Reef, which extends more than 1,300 miles along the continent's Pacific Coast. This is only a third the length of the Great Wall of China, but impressive enough when you see its massive bulk—80,900 square miles—and consider its 400 species of coral and 1,500 species of fish.

14. 1–c
    2–b
    3–f
    4–g

5–c and d, i.e. Ghana and southeast Florida

6–a

7–e

8–h

Ivory Coast is a country in West Africa. Barbary Coast is the southwestern shore of the Mediterranean, the coast where the U.S. Marines found themselves on "the shores of Tripoli" while pursuing Barbary pirates. The Skeleton Coast is the desert coast of Namibia, formerly South-West Africa. The Tanga Coast is very far from where you would see Brazilian bikinis (which are called *tangas*); it is the Arab-settled coast of Kenya and Tanzania. The Gold Coast was the British colonial name for Ghana; it is also a chamber-of-commerce term for the eastern coast of lower Florida. The Redneck Riviera is a string of honky-tonk bars and amusements that extends from Panama City, Florida, to Gulfport, Mississippi, along the Gulf of Mexico. The Riviera is on the coast of Italy and France, where it is called the Côte d'Azur. The Mosquito Coast, home of the Mosquito Indians and many mosquitos, is on the Caribbean side of Central America and is both setting and title of a novel by Paul Theroux.

15. **a** Off the New England coast
   b Off the Newfoundland coast
   c Home office: London, England

16. Strait answers:

					16–c
1–r	**4**–s	**7**–n	**10**–k	**13**–h	**17**–a
2–q	**5**–o	**8**–m	**11**–j	**14**–g	**18**–e
3–p	**6**–b	**9**–l	**12**–i	**15**–f	**19**–d

17. These currents all flow along the west coasts of continents toward the equator. They bring cool, up-welling nutrient-filled water near the coasts, and thus create the world's greatest fisheries.

18. Norfolk Island—for Australia.

19. The *Bounty's* mission was to bring breadfruit for planting in Jamaica to feed slaves.

20. Greenland was "discovered" in 986 by the Norseman Eric the Red, who presumably might have named it Redland but refrained. At 840,000 square miles, it is the world's largest. We place the word *discovered* in quotes because Greenland did not need discovery, having been inhabited comfortably for several thousand years by the Inuit, whom we presumed to rename the Esquimau or Eskimo. (Granted, the Inuit, being human, were also presumptuous—the name meant "the People.") The Norse founded two settlements on Greenland's west coast, and at one point three thousand Norse lived there. But because the climate kept on getting worse, by 1500 they had all moved elsewhere or died.

21. You can't tell a big island from a continent except by the name; it is purely arbitrary to designate one big island as a continent and another big island as just another big island. Blame geographers.

22. Baffin Island, 183,810 square miles, the world's fifth largest, is Canada's biggest.

23. New Guinea, 306,000 square miles, is the largest island in the Pacific.

24. New Guinea is the world's second largest island, after Greenland.

25. Islands are formed in four ways. Some are chunks of continents lost at sea. Greenland is just a small continent, and Fiji is another, even smaller one. The Channel Islands off California are part of North America. Others, the offshore barrier islands such as Cape Hatteras, are built up from beach sand and mud from nearby rivers. Some islands—Hawaii, Tahiti, Iceland—are old volcanoes. And others are made of limestone, built up from coral reefs and atolls.

26. Certainly; large parts of the tropical oceans have desert climates.

27. Long Island is a terminal moraine, a limestone ridge covered by a pile of mud, gravel, and sand left behind by a glacier during the last ice age. It is not a bad place to live, although the railroad could be better.

28. Yes and no. England obviously did; Japan did not. Evidently the environment offers opportunities, but cultural patterns determine whether they are seized.

29. The Puerto Rico Trench, northwest of that island— 28,374 feet.

30. Not Don Ho and not Arthur Godfrey. It was Portuguese sailors from the island of Madeira. They called it the *machete* then, no relation to the big cane-chopping knife, but Hawaiians renamed it. In Hawaiian, *ukulele* means "little jumping flea."

31. The Galápagos Islands.

32. The islands listed are claimed by the following countries:

   a Chile
   b United Kingdom
   c South Africa
   d Australia
   e Norway
   f United Kingdom
   g France
   h Vanuatu
   i United States
   j United Kingdom
   k United Kingdom
   l P.D.R. of Yemen
   m United Kingdom
   n Italy
   o United States

33. Bora-Bora, says the author of *Tales of the South Pacific*.

34. Cuba, which he saw in 1492.

35. In Japanese, *shima* and *jima* mean "island."

36. Bermuda is in the Sargasso Sea.

37. Islomania is an irresistible attraction for islands. Most of us seem possessed by it to one degree or another, whether we daydream of hacking out a new world on the Mosquito Coast with a machete or escaping to a free-market paradise like the Caymans.

38.

1–h	5–d
2–g	6–b
3–f	7–a
4–e	8–c

39. In 1964, Leicester Hemingway, Ernest's younger brother, who until then was known mainly as the promotion man for Florida jai alai and as Ernest's younger brother, announced the birth of a republic. He called it New Atlantis. Hemingway's new nation was brought forth on a formerly submerged sandbar six and a half miles off the southwest coast of Jamaica. It was the smallest nation ever conceived: population, one part-time inhabitant; area, eight square yards; elevation, three inches above sea level when the sea was calm. Even so, it cost him some $50,000 to build New Atlantis, Hemingway claimed, mostly for the hauling of rock and soil.

The exact location of this improbable republic was a state secret, Hemingway said coyly. But he issued a declaration of independence, drew up a constitution, and notified the United Nations and the rest of the world: "We are a peace-loving nation."

Hemingway said he was not worried about losing his American citizenship. He relied upon a century-old law called the Guano Act, which permitted, he said, U.S. citizens to claim any uninhabited island in international waters so long as it contained deposits of guano—which, of course, is bird dung highly valued as fertilizer for its high concentration of urea. And there was indeed plenty of it on the island of New Atlantis. Hemingway imported it himself.

40. These are Galápagos Islands.

41.

1–h	5–b
2–e	6–d
3–g	7–c
4–f	8–a

42.

1–f	4–a
2–e	5–b
3–d	6–c

43. They are called abyssal plains.

44. Allentown claims to have the world's largest indoor wave machine and not long ago began holding professional surfing competitions there.

45. "Chicken."

CLIMATE

The difference between weather and climate is a little like the difference between tactics and strategy. Weather is what the atmosphere is doing on a daily basis, usually in your neighborhood. And climate is the long-term average condition of the atmosphere. We may talk about weather and not do anything about it, but weather doesn't treat us with similar courtesy. It influences us:

our choice of clothing, our energy bills, our health.

Our behavior? Perhaps it's true that more soft drinks are bought on windy days and that domestic violence increases in Los Angeles when the Santa Ana winds blow, but evidence for this intriguing question remains ambiguous.

1. Why does the weather in most places change every few days?

2. If you think the earth is heated by the sun—and you're correct if you do—why is the air warmer near the ground than closer to the sun?

3. How hot can it get in the shade?

4. Why is it hotter in the tropics than near the poles?

5. What part of the world would have the highest temperatures in July?

6. Why are there deserts?

7. What is unique about the sky above a desert?

8. Where would you guess is the world's coldest capital city? Where is the next coldest?

9. What is today's temperature in Quito, Ecuador?

10. Why are winters warmer in London, England, than in London, Ontario, even though Ontario's London is nearly six hundred miles farther south?

11. Take a deep breath. What gas amounts to 78 percent of air we breathe?

12. Why are there seasons?

13. What is a more common name for a low-pressure cell?

14. What is a high-pressure cell?

15. What are westerlies?

16. What are the trade winds?

17. What are the horse latitudes?

18. What are the doldrums?

19. Southern California is famous for its climate. What do you call this type of climate?

20. Where are . . .
   a The hottest average daily temperature on earth?
   b The coldest average daily temperature on earth?
   c The coldest recorded temperature where people actually live (if you can call it living)?
   d The coldest temperature ever recorded anywhere on earth?

21. How can you tell the difference between a wet climate and a dry climate?

22. What causes monsoons?

23. Where would you expect the most solar energy to reach the ground:
   a A desert
   b A tropical rain forest
   c A northern coniferous forest

24. What is fog?

25. What do you call the puffy little clouds you see in the blue sky of a nice day?

26. In which state does it rain the most?

27. What part of the United States has the most thunderstorms?

28. What weather phenomenon kills the most people, and where in the United States do you find it the most?

29. Does lightning ever strike twice in the same place?

30. **a** What country has the most tornadoes?
   b What part of the United States has the most tornadoes?
   c What are tornadoes called in Iowa?

31. Where do hurricanes come from?

32. What are hurricanes called in the Indian Ocean?

33. What are hurricanes called in East Asia?

34. What is
   a A *foehn*?
   b A chinook?

35. What country has the most windmills?

36. How does the climate in a city differ from the climate on an adjacent farm?

37. Where do the farmers *not* want rain during the summer, and why?

38. If it's cold where you are in January and you'd like a hot vacation, where's the best place in the world to go?

CLIMATE

1. Weather changes for various reasons, the main one being the local, regional, or global motion of the atmosphere. Local effects include the trapping of polluted air beneath layers of warm air in basins such as Los Angeles and Mexico City. These are called *inversions*. Most of the changes we see are brought on by regional movement of large packages of air called *air masses*. Air masses pick up the characteristics of the place where they were formed and carry them along to other places, where they may be a surprise. Air masses from the Gulf of Mexico are warm and wet. Air masses from Siberia are cold and dry. Air masses move because they are dragged along by the *jet streams*—high-speed winds, far up in the atmosphere, that blow constantly. The jet streams are caused by a combination of the earth's spinning and the temperature contrast between the cold polar and hot equatorial regions. Thus the atmosphere churns, constantly delivering new weather so that we'll have something to talk about every day.

2. The air is warmer close to the ground because most of the sun's energy passes through the atmosphere without stopping, and where it finally does stop is on the ground, the rooftops, and our car seats. It warms these surfaces, which then radiate warmth back into the surrounding air. The higher you go, the cooler it gets.

3. It can get pretty hot in the shade. The record is 136.4 degrees F, in the Sahara, in Libya. Where no river exists and rain may not fall for years, it is really hot. In North America, the hottest recorded temperature was at Death Valley: 134 degrees.

4. It is hotter in the tropics, or near the equator, because the sun's warming rays strike the earth straight-on. Farther north or south, they strike glancingly because of the earth's curved surface, so that the same amount of heat has to spread over more ground. Also, at high latitudes, the sun's rays lose some of their energy just traveling through a thicker layer of atmosphere.

5. The highest July temperature would be found in the western Sahara, in Libya or Algeria, where the summer average is over 100 degrees.

6. There are deserts because there are places where it does not rain very much. Actually, there are two reasons why regions have dry climates. A region may be in the interior of a continent, such as in Central Asia or western North America, far from the main moisture sources, the oceans. Some of these are in "rain shadows," where they are blocked from moist air by upwind mountains. Secondly, deserts are common in those parts of the tropics where air is usually descending, as in the Sahara and Middle East. Rain only happens when air ascends, is cooled, water condenses, and becomes raindrops.

7. Desert skies generally have no clouds. There is little moisture to make them. This also means there will be little shade during the hot days, and no insulating blanket of clouds to keep the warmth in at night. This is why temperatures fluctuate so widely in the desert, from record-setting heat in the daytime to sometimes below freezing at night.

8. The coldest capital city is Ottawa, Canada, where the average daily high temperature in January is minus 6 degrees C. Second is Ulaanbaatar, Mongolia. Moscow is third.

9. The temperature in Quito today is probably 69, 70, or 71 degrees F. This is interesting because temperatures vary widely in most places on earth, and Quito is one of those unusual spots where it does

not. The explanation: Quito is 9,300 feet high and only eight miles south of the equator. High-altitude equatorial settings have much the same weather every day.

10. Western Europe is warmer because it is downwind from the Atlantic Ocean, whereas southern Ontario is downwind from the interior of North America. Oceans are slow to cool, and they never get as cold as continental interiors. These features influence climate. London has a second advantage: its proximity to the Gulf Stream, which brings up warm water from the Caribbean.

11. Nitrogen. Only 21 percent is oxygen.

12. There are seasons because the earth, as it moves in its elliptical orbit around the sun, exposes itself to the sun differently at different times of the year. The earth's axis stays at the same tilt—pointed out into space at the same angle—all year, which is why the North Star is in the north whether the season is winter or summer. But as the earth moves about the sun, the Northern Hemisphere is more exposed to the sun during one period of the orbit, giving that hemisphere its summer, and then the Southern Hemisphere is exposed during the other period of the orbit, getting its place in the sun.

13. A low-pressure cell is a cyclone. This is a volume of ascending, converging air. Since the air is rising, a barometer will read low pressure. In the Northern Hemisphere, cyclones turn counterclockwise; in the Southern, they turn clockwise. Cyclones are important to people because they often lead to rain. The only way to make rain is for wet air to cool, and the easiest way for this to happen is for something to make it rise to cooler environs aloft, the way a cyclone does. Humid regions often have cyclones passing over them.

14. A high-pressure cell is called an *anticyclone* and is a volume of air descending to the earth and diverging. It will read high pressure on a barometer. Anticyclones turn clockwise in the Northern Hemisphere, counterclockwise in the Southern. There is little chance of rain in the vicinity of an anticyclone, because it is essentially cooler air descending into a warm area. You frequently find these over deserts.

15. The westerlies are winds from the west that churn around the planet in the middle latitudes (forty to sixty degrees latitude). Very far south, where there is little land to interfere, the westerlies are very persistent, and old mariners called these latitudes the "roaring forties," "furious fifties," and "screaming sixties." Albatrosses can circumnavigate the globe by riding on these winds.

16. The trade winds are steady northeast winds in the tropics of the Northern Hemisphere, and steady southeast winds in the tropics of the Southern Hemisphere. They blow toward the Equator, and for centuries provided dependable wind power for sailing ships—hence the name. (Winds are always named for the direction they blow *from*.)

17. The horse latitudes, in the North Atlantic south of Bermuda, are often a place of quiet, stable weather, of light winds and calms. In the days of Spanish colonization of the Americas, passage through here was often unduly prolonged by these calms, and to lighten ship and avoid having to feed extra mounts, the animals were often jettisoned. They floated for a time.

18. Doldrums are isolated spots of calm within the equatorial low-pressure zone. In low-pressure areas here the air is always rising, instead of circling in a great spiral the way it does elsewhere.

19. The wonderful climate of southern California is called, of all things, "Mediterranean," because it is like the climate surrounding that sea. This means wet winters and dry, sunny summers. Low-pressure cells travel along the westerlies and cause the wet winters. High-pressure cells dominate in the summer, bringing warmth. Areas with this Mediterranean climate comprise a tiny proportion of the earth's land surface, which has a great deal to do with the high price of real estate within them.

20. a The hottest daily average is 94 degrees F, in Dallol, Ethiopia.
   b The coldest daily average is minus 72 degrees F, at Nedostupnosti, a Soviet base on the South Polar ice dome in Antarctica.
   c The coldest temperature ever recorded where people live, if you can call it living, is minus 96 degrees F, at Oymyakon, Siberia.
   d The coldest recorded temperature anywhere on earth, ever, was minus 126.9 degrees F, at another Soviet base in Antarctica, Vostok.

21. A wet climate, scientists say, has more rain than evaporation. A dry climate has more potential for evaporation than it has rain. In humid climates, the surplus water makes rivers, which help determine how the land looks. In arid regions, there is not enough rain to make permanent rivers.

22. Monsoons are seasonal alterations in wind patterns over large continental areas. Over Asia, the winter brings colder temperatures, sinking air, and high pressure. Winds flow outward, bringing dry, cooler weather to the continent's edges—places like India and Vietnam. In summer, the land heats up, low pressure develops in the interior, and wet air comes

inland from the surrounding oceans. The summer monsoon brings rain to India and Southeast Asia. The same principle operates on all continents.

23. Sunlight is most intense in the equatorial regions, but it is frequently cloudy there, and the sunlight is reflected away. The most solar heat hits the ground in the tropical deserts, such as the Sahara and central Australia.

24. Fog is simply a cloud sitting on the ground. A cloud is a parcel of air in which water condensation is underway. Condensation droplets are tiny—several thousand times smaller than a raindrop.

25. The puffy little clouds you see on nice days are called cumulus clouds. They, like the other clouds, were named by an Englishman named Luke Howard in 1804.

26. In Hawaii, on Mount Waialeale, which gets 460 inches of rain per year. A side of this mountain faces into the prevailing winds. As the wind moves upslope, the air cools, producing clouds and eventually rain. This is called *orographic rainfall*, and it happens to a greater or lesser degree on most mountains.

27. The Midwest has forty to fifty thunderstorms a year, but the Florida peninsula has twice as many.

28. Lightning kills two hundred people a year, and it's worst in central Florida. More lightning strikes there than anyplace else in the United States, some say the world. Lightning, researchers say, is attracted to specific geographic points and to tall, exposed objects. In Florida, it is most likely to strike in spots where sea breezes hit coastal headlands. Hurricanes and tornadoes follow close behind lightning in hazardousness. They kill about 190 a year.

29. There's no reason in the world why lightning won't strike twice in the same place. In fact, two scientists from the U.S. Department of Commerce watched 120,000 lightning bolts in Colorado one recent summer, and found that two, or more, strikes in the same spot are not at all uncommon, and that one particular ridge near the town of Castle Rock was always getting hit by bolts from the blue. The most dangerous places to be during electrical storms are on the water, on tractors, under trees, in open fields, or on golf courses. That's why Lee Trevino always carries a three iron on the golf course now. He was once hit by lightning, nearly died, and doesn't want it to happen again. He says: "Not even God can hit a three iron."

30. a The United States, reporting more than seven hundred a year, has the most, followed by Australia and then by India, according to Alvin Samet of the National Hurricane Center in Miami.

b Oklahoma, particularly between Tulsa and Oklahoma City. This is an area where cool, dry air from Canada often collides with warm-wet air from the Gulf of Mexico, creating classic conditions to spawn tornadoes.

c Cyclones.

31. Hurricanes begin as low-pressure cells in the trade-winds latitudes. Water condenses from the wet air, freeing latent heat energy, which is converted into wind speed, accelerating the spinning wind patterns to hurricane speeds—seventy-five knots or more. Once it has developed, a hurricane can wander around the tropics for days or even weeks, until it strikes a continent or collapses over cool ocean waters.

32. Cyclones.

33. Typhoons.

34. a *Foehn* is the name for hot, dry, gusty winds that blow down the Alps into northern Europe during certain times. They correlate strongly, though no one is certain why, with increases in medical complaints, accidents, crimes, and suicides.

b A chinook is the same kind of wind east of the Rocky Mountains.

35. No one seems to keep track of these important matters, but the United States is a good bet. Over California's Altamont Pass alone, more than 2,200 windmills pull electrical energy from the open skies. The Netherlands has only 1,000 working windmills, although they are far more picturesque.

36. The city will be hotter (because the city generates and holds heat), dustier (with air pollution), and foggier (humidity is higher, and the dust provides more grains for condensation to form around). There will also be more rain in the city because the heat forces air to rise, sometimes cooling it enough to produce rain.

37. Farmers in the San Joaquin Valley of California raise raisins. Most raisins begin life as Thompson seedless grapes, which are picked when ripe and then simply placed on sheets of paper between the grape rows. For three weeks, they lie there turning into raisins. If it rained, they'd turn into something worse. The farmers picked this place to grow raisins with irrigation precisely because it doesn't rain in the summertime, and they could—in a very limited way, just enough to make raisins—control the weather.

38. If you're in the Northern Hemisphere and suffering a cold winter January, the best place to get warm is in the Southern Hemisphere, where it's summer. The average temperature in northwestern Australia in January is over ninety degrees.

ENVIRONMENT

A little bromeliad plant sitting in the high crotch of a tree in the Amazonian rain forest, with tadpoles and waterbugs swimming around in the teacup of water forming a lagoon among its leaves, is an environment. It is a unit that has boundaries and components with something in common inside. An environment is the sum total of the things that give a place its uniqueness, that allow it to be defined as a place, in and of itself—a place in which everything inside has a relationship with every other thing inside. It is the relationships that hold things together and that keep the environment in equilibrium. The earth itself, contrasted with space, is an environment like that.

1. Where do new forms of life come from?

2. Why don't all plants and animals live everywhere?

3. Does the number of species vary with the kind of environment?

4. Following are five things humans do, each with the same environmental significance. What is it?
   a Harvest wild plants and animals, as in fishing or gathering nuts
   b Cultivate domesticated plants and animals, as in cornfields and feedlots
   c Eliminate competing organisms, as in weeding or shooting wolves
   d Inadvertently or deliberately disperse or increase other organisms, as in importing cockroaches concealed in TVs from Japan or shipping parrots for pets
   e Disrupt or destroy natural habitats, as by deforestation, overgrazing, oil drilling, condominium building

5. We're all stuck with life in the food chain and have to earn a living off the land. But there are ways to save energy in the process—even in what we eat. What is the most obvious way?

6. Where does the oxygen in the atmosphere come from?

7. Can you match these national parks with the states in which they are found?

   1 Bryce Canyon   **a** Arizona
   2 Everglades   **b** Colorado
   3 Yosemite   **c** Montana
   4 Crater Lake   **d** Washington
   5 Mount Ranier   **e** Oregon
   6 Glacier   **f** California
   7 Rocky Mountain   **g** Florida
   8 Grand Canyon   **h** Utah

8 . There are deserts and deserts. Can you describe some different types?

9. The bandicoot in Australia's outback, the jackrabbit in the North American desert, the hedgehog in the Gobi, and the fennec in the Sahara all have something in common—What? And why?

10. Match the desert with its country:

1 Gibson Desert — a United States
2 Kalahari — b Chile
3 Takla Makan — c Algeria
4 Thar Desert — d Egypt
5 Peski Karakumy — e Saudi Arabia
6 Rub 'al-Khali — f USSR
7 Arabian Desert — g India
8 Erg Chech — h China
9 Atacama Desert — i Botswana
10 Mojave Desert — j Australia

11. There is an unusual kind of desert, most examples of which are found in central Asia. For living things that are not properly adapted to it, this kind of desert has two hazards—the usual one of aridity, and another one. What is it?

12. There are two ice caps on earth. Where are they?

13. Tropical savannas have rainy summers and dry winters, but are warm all year. The vegetation is usually some sort of open forest with scattered grasses and shrubs. The largest areas of savanna are on two continents. Can you name them?

14. Three regions of the world were once covered by great hardwood forests. Where are these three regions, and what happened to the hardwood forests?

15. Species diversity is the number of plant or animal species per unit area. How does species diversity change as you go up a mountain?

16. How does species diversity change as you travel from the equator to the poles?

17. Continents are one kind of environment. Islands are another. How do these differ in the number of living things that inhabit them?

18. Some places produce lots more living matter than others—they teem with life. The measure of this is called *net biological productivity*: the amount of organic matter produced per square meter per year. What sorts of environments would you say have the highest biological productivity?

19. Unless you are a very observant traveler indeed, you may have missed one of the most important features of environmental geography in the places you've visited: the soil. See if you noticed in your travels what sort of environment you'd find these soils in:

a Deep black soil with lots of decaying organic matter
b Deep red clay with surprisingly little natural fertility, considering the abundant vegetation it seems to support
c Black surface layer with a gray layer beneath

20. A single family of plant life covers about a quarter of the earth's land. What family is this?

21. What country emits the most sulphur dioxide air pollution?

22. What is America's greatest conservation problem?

23. The last three great stands of humid tropical forest are situated where?

24. There are fewer songbirds in the United States every year, according to biologists. Why?

25. What is a moor?

ENVIRONMENT

1. They evolve. To put it one way: Novelty derives from random and imprecise changes in the transfer of genetic information from parent to offspring. Or to put it another: Changes are largely accidental. Offspring are made up of a little from this parent and a little from that, plus a little something new from chance.

This can produce some pretty weird items: plants that eat animals, lizards ten feet long, creatures that hang upside down in trees all their lives turning green with algae and descending only to feed the tree with their droppings.

Then natural selection takes over. Success in the environment is what determines whether a sloth will live on and reproduce itself. Some breeding groups are isolated—as on the Galápagos Islands, where Darwin first sorted out the idea of evolution. This isolation limits genetic diversity and localizes the population to a particular environment, providing the opportunity for natural selection to establish new characteristics.

2. Plants and animals have come to be what they are precisely because of the environment they live in, so they can't very well live just anywhere. This is called *adaptation*—living things evolving with life processes and ways of making a living geared to particular environments. Since environments vary because of the earth's terrains and climates, there is no way that every plant and animal can be adapted to living just anywhere. Obviously, the more specialized a given creature is, the narrower the range of possible environments in which it can succeed.

3. Yes—generally, there are more species in warmer environments. The humid tropical forests have more species than anywhere else:

Alaska has fifteen native bird species, and one bat species.

Michigan has thirty-five native bird species, and seven bat species.

Panama has seventy native bird species, and thirty-one bat species.

4. These are the five major ways we humans affect the survival or failure of other species on the planet. You will notice that most are to other species' detriment, without necessarily improving our own condition.

5. Most humans are omnivores, eating both plants and animals. When you eat bread, you take one step, or one link of the food chain. If you eat beef, the chain is two links long: from the cow to the grasses it consumed, to you. If you fried and ate a bald eagle, then you'd be eating bird that ate predator fish that ate herbivore fish that ate aquatic plants. Because energy is lost at each link in the chain, humans can conserve energy largely by consuming plants.

6. Every green plant—the redwood tree, the geranium on the windowsill—combines water and carbon dioxide with sunlight in a process called photosynthesis and gives off pure and life-giving oxygen.

7.
1–h	5–d
2–g	6–c
3–f	7–b
4–e	8–a

8. Deserts certainly are not all the same. In North America alone, there are five major deserts, each different from the next. While deserts are defined largely by their lack of vegetation, ironically they differ mainly in the kind of vegetation they do have. In the Great Basin, the biggest and bleakest, extending across much of the country between the Sierra and the Rockies, the plant you see most often is sagebrush. In southeastern California is the Mojave, home of the Joshua tree. South of this is the Sonoran desert, with its giant, dramatic saguaro cactus; and farther east is Mexico's Chihuahuan Desert, studded with spiky agave plants. The Painted Desert of northern Arizona and New Mexico is among the most barren, but is covered with the bright red-orange trunks of petrified trees.

9. These desert-dwelling animals all have enormous ears with lots of blood vessels near the surface of the skin, so that any breeze blows heat away and cools their blood.

10.
1–j	6–e
2–i	7–d
3–h	8–c
4–g	9–b
5–f	10–a

11. This unusual kind of desert is called a "cold desert," and its second hazard is bitter winters.

12. The world's two ice caps are in Greenland and Antarctica. The ice there may be more than ten thousand feet thick.

13. The largest areas of tropical savanna are in South America and Africa, both north and south of the equator.

14. The three once-wooded regions are eastern China, which has scarcely any natural forest left (the government has begun a tree-planting program); western Europe, where natural forest survived in some abundance until the Middle Ages; and eastern North America, where the forests survived until the late 1800s. All were felled by humans, for agriculture, to build shelters and ships, for fuel. In Europe, great expanses were also blasted by two wars.

During the last thirty years, technical progress in agriculture has made it possible—at least in the developed countries—to grow more on less land. As a result, hardwood forests are returning to eastern North America and western Europe, just in time to be threatened by something new: acid rain.

15. Species diversity decreases as you go up mountain slopes. This is most visible in the tropics, where the number of tree species, for instance, dwindles from hundreds per square mile near sea level to only a few at the edge of the tree line.

16. The number of species decreases the farther you get from the warm equator and nearer you get to the cold poles.

17. Islands tend to have fewer species (per unit of area) than continents. The farther an island is from the continents, and the smaller it is, the fewer species it has. The greater the distance, the more difficult it is to colonize across that distance.

18. Humid tropical forests and freshwater marshes score very high in biological productivity, each producing about three times as much as prairie grasslands. Deserts and ice caps, as you guessed, have the lowest productivity.

19. a The deep black soil is found in prairie grasslands.
b The deep red infertile soil is found in humid tropical forests, where almost all the nutrients are locked into the growing vegetation.
c The black soil with gray underneath is found in the cold coniferous forests.

20. Grasslands cover a quarter of the earth's surface.

21. The United States emits 20 percent of the world's total of sulphur dioxide, which combines with water to make sulphuric acid and acid rain. The Soviet Union, emitting 18 percent, is trying hard to catch up.

22. Obviously, this is debatable, but our greatest problem may well be soil erosion. The problem embodies no conflict, like the clubbing of harp seal pups or catching porpoises in the nets of tuna fishermen, which may be why it has received less media attention. But soil erosion is much more serious today than it was in the Dust Bowl 1930s, and it's worsening all the time. What is being lost is the upper "horizon," as layers of soil are called. This is the A-horizon, where plants have most of their roots and find most of their nutrients. Where it is severe, loss of the A-horizon has ended farming, perhaps forever. It often takes thousands of years to form an A-horizon.

23. The last remaining humid tropical rain forests are in the Amazon Basin, in the Congo Basin (in Zaire), and on various islands of Indonesia, particularly Borneo, Celebes, and nearby Malaysia and New Guinea.

24. Biologists are divided on why fewer warblers, thrushes, and tanagers are found in the United States these days. But they agree on two problems. U.S. Fish and Wildlife biologist Marshall Howe says that habitat destruction inside the United States is harming warblers, vireos, and flycatchers. Another problem, according to Sandy Sprunt of the National Audubon Society, is destruction of tropical forests in Mexico and Central America, where migratory birds have always found winter retreat. "Both are a problem," Sprunt says, "But the destruction in the United States is not going on nearly as fast." The tropical forests are being bulldozed to make way for cattle ranches to grow fast-food hamburgers for North Americans—including, no doubt, some bird lovers.

25. A *moor* is an open, rolling wasteland, sometimes boggy and often covered with heather.

PLANTS

Plants are thought to be the living entities that don't move. But don't depend on it; they've devised some innovative styles of long-distance sex and long-range seed dispersal.

With 300,000 species of green plants and fungi alive, studying them is not a simple matter.

1. Five sources of caffeine are found in nature. One is coffee. How many others can you name?

2. If you dropped a pine needle in a northern coniferous forest and a leaf into an equatorial rain forest, which would last longer?

3. What is "forest death"?

4. Why is the light actually green in the Olympic forest of Washington?

5. Plants living in the large climatic regions of the world have adapted to their environments. Generally, there are eight types of vegetation zones. Match each with its definition:

1 Found around the Arctic Circle and sub-Antarctic islands, this has small shrubs, stunted trees, many sedges, and vistas of summer wildflowers. Summers here are short and cool.

    a Desert
    b Temperate grasslands
    c Tropical rain forest
    d Temperate forest
    e Tundra
    f Northern coniferous forest
    g Tropical savanna
    h Chaparral

2 These great stands of fir, spruce, larch, and pine extend across the north of Canada and Eurasia. Summers are warm, winters severe.

3 Great stands of deciduous, broad-leafed trees growing in humid middle latitudes, most now gone and replaced by farms, pastures, and suburbs.

4 In equatorial regions of nearly daily rainfall are many species of trees, lianas, and epiphytes—air plants.

5 In areas too dry for forest but wetter than deserts are these great grasslands, like the Great Plains of North America and the grasslands extending from Poland to western Siberia.

6 Grassy areas with trees and shrubs, as are found across much of Africa and tropical South America.

7 California's name for the shrubby areas produced by its summer-drought Mediterranean climate; they can be seen as background in many television series.

8 Areas of drought, which produce limited but hardy vegetation.

6. Match the region with the name of its unique grassland:

1 Steppes   **a** Northern South America
2 Pampas   **b** Central Asia
3 Llanos   **c** Southern South America
4 Prairies   **d** North America
5 Grassveld   **e** Southern Africa

7. What kind of vegetation is partly responsible for the development and persistence of Moscow as a city?

8. Which sort of plant has the biggest seed, and where is it found?

9. There are two kinds of tropical coastlines. What are they, and which kind of plant dominates on each?

10. Where do the oldest trees on earth grow?

11. Which is the most widely planted agricultural tree?

12. Where would you find Venus flytraps in the wild?

13. Where were these vegetables found in the wild, before they were domesticated by humans?

1 Peas   **a** India
2 Eggplant   **b** Southwest Asia
3 Beets   **c** Southeast Asia
4 Cabbage   **d** South Asia
5 Asparagus   **e** Central America
6 Pumpkin, squash   **f** East Africa
7 Cucumbers   **g** West Africa
8 Tomatoes   **h** The Mediterranean
9 Spinach   **i** China
10 Onions   **j** The Andes
   k Eastern South America

14. And these fruits?

1 Strawberry   **a** India
2 Pineapple   **b** Southwest Asia
3 Papaya   **c** Southeast Asia
4 Plum   **d** South Asia
5 Peach   **e** Central America
6 Citrus   **f** East Africa
7 Avocado   **g** West Africa
8 Grape   **h** The Mediterranean
9 Apricot   **i** China
10 Muskmelon   **j** The Andes
11 Kiwi   **k** Eastern South America

15. These nuts?

1 Cashew   **a** India
2 Almond   **b** Southwest Asia
3 Walnut   **c** Southeast Asia
4 Pistachio   **d** South Asia
5 Peanut   **e** Central America
6 Brazil nut   **f** East Africa
7 Water chestnut   **g** West Africa
8 Kola nut   **h** The Mediterranean
   i China
   j The Andes
   k Eastern South America

16. And these grains?

1 Maize (corn)   **a** India
2 Wheat   **b** Southwest Asia
3 Oats   **c** Southeast Asia
4 Barley   **d** South Asia
5 Grain sorghum   **e** Central America
6 Soybean   **f** East Africa
7 Rice   **g** West Africa
8 Buckwheat   **h** The Mediterranean
   i China
   j The Andes
   k Eastern South America

17. What plant, if any, grows almost everywhere in the world?

18. Grasslands with oak trees are the natural vegetation of much of the California coast and valleys. The Forty-niners found them when they arrived to prospect for gold. Where did these grasses come from?

19. Can you name a fruit you probably eat at least weekly, perhaps even daily, but that only a hundred years ago was grown only as an ornamental?

20. Humans grow three different strains of the hemp plant, each for a different use. What are these uses?

21. Why is it that you can tear off a corner from a Burpee seed packet, pour the seeds into a hole in the ground, water them from time to time, and end up with lettuce or radishes, while you can scarcely get a wild plant—even some that volunteer all over your yard—to grow if you try?

22. About how long has it been since humankind domesticated a significant new food crop?
One hundred years
Five hundred years
Two thousand years
Five thousand years

23. It is a curiosity of U.S. agriculture that the crops we grow came from somewhere else: wheat from Asia, soybeans from the Orient, maize from the New World tropics. But there is another crop that originated here and is now grown commercially in southwestern Europe. Can you name it?

24. In Mexico, where it is grown, what is *maíz reventador?*

25. Two members of the heath family native to New England are among that region's best known products. What are they?

26. New Jersey is the most densely populated state in the union, yet in its center is a region where almost no one lives. It is named for its predominant species of tree and has had the same evocative name since the seventeenth century. What is it?

27. Remembering the Alamo is something they do in Texas, where the Alamo is. *Alamo* means cottonwood tree in Spanish. So, what sort of trees have been planted along San Antonio's Riverwalk, the tourist district of shops and hotels along the San Antonio River near the Alamo?

PLANTS

1. Caffeine is found in coffee, tea, yerba maté, cola plant, and cacao. Every one was discovered by primitive humans, and every one was used the same way we use them: to prevent fatigue. And science has not discovered another source since.

2. The pine needle might take five to seven years to decay, but the leaf in the rain forest rots away in about six weeks.

3. *Forest death* is the name given, in Europe, to mysterious stalled growth in trees. In West Germany, for example, this has afflicted a third of the famous forests since 1979, at a cost to the lumber industry of $200 million a year. It is linked to auto exhaust and acid rain.

4. This forest is rain forest, "dominated by water," in the phrase of one naturalist, and so humid that droplets of water hanging in the air reflect the greenery all around giving the air a green tint.

5. 1–e   5–b
   2–f   6–g
   3–d   7–h
   4–c   8–a

6. 1–b
   2–c
   3–a
   4–d
   5–e

7. Great forests of coniferous trees stretching across what is now the USSR protected Moscow from marauding Mongols on horseback, who preferred to fight in open spaces. These vast forests are called the *taiga*, and not surprisingly, they figure powerfully in the Russian literature and self-image.

8. Logic suggests the biggest plant might have the biggest seed, so you'd look among the trees, and that would be wise. But the biggest trees—the sequoias—have tiny seeds. The really big seeds are found on some of the smallest, slenderest of trees, the palms. One of them, found only in the Seychelles Islands in the Indian Ocean 850 miles off the coast of Africa and called by the French the *coco de mer* ("palm of the sea"), has a seed bigger than a basketball. It weighs 50 pounds.

9. Two types of tropical coastlines are "muddy shoal" and "sandy beach." The former is dominated by mangroves, trees that have mastered the clever trick of living in salty or brackish water yet have freshwater metabolism. The tropical sandy coasts have coconut palms, whose tough fibrous seed husk allows the coconut to float across oceans and sprout in a new place.

10. The oldest trees on earth are the wind-sculpted bristlecone pines, which live high in the White Mountains of eastern California and Nevada. One of these trees may be five thousand years old.

11. The coconut palm exists across all the tropics and subtropics, and its wood, shells, fibrous husks, fronds, meat, and milk are used in many ways. Though seen throughout south Florida and southern California, it is native to neither of those places.

12. Venus flytraps live in the forests of the southeastern United States and are only one of forty-five meat-eating plant species in this country. The reason why they are carnivorous is a mystery, however, since they have green leaves that photosynthesize and presumably would not need to catch bugs and small amphibians for food. One explanation: The diet of meat provides trace minerals not available from the soil.

13. 1–b   6–e and j
    2–a   7–f
    3–b   8–e and k
    4–i   9–b
    5–h   10–b

14. 1–j   5–b   9–i
    2–k   6–c   10–e
    3–j   7–e   11–i
    4–i   8–h

15.

1–k	5–k
2–d	6–j
3–b	7–c
4–b	8–f

16.

1–e	5–f
2–b	6–i
3–b	7–f and d
4–b	8–i

17. The closest thing to a ubiquitous plant is *Plantago major*, a small plant with a rosette of broad leaves that clings near to the ground. It occurs on every continent except Antarctica. It is a "weed" spread by people and is commonly known as plantain.

18. The grasses now growing along the California coast, although they predated the Forty-niners, are far from native. They come from the Mediterranean area and were brought here accidentally by the earliest Spanish explorers. Once in California, these Old World grasses, including wild oats, prospered and so largely replaced the true native grasses.

"The plants which are growing unasked and unwanted on the edge of Santa Barbara," writes botanist Edgar Anderson in *Plants, Man and Life*, "are the same kinds of plants the Greeks walked through when they laid seige to Troy. Many of the weeds which spring up untended in the wastelands where movie sets are stored are the weeds which cover the ruins of Carthage and which American soldiers camped in and fought in during the North African campaign. . . . Some of them had evolved through a whole series of civilizations, spreading along with man from the valley of the Indus to Mesopotamia and on to Egypt and Greece and Rome."

They came here, no doubt, in the fur of livestock and in the hay and seeds of other field crops.

19. The tomato, introduced to Europe after the Spanish conquest of the New World, was grown mostly as a curiosity for generations. It belongs to a Central and South American family of weedy little shrubs with little red or orange berries. When Europeans brought it to North America, it was rejected here, too, at first. Some people even decided it was poisonous, and called it the "love apple." The Italians first saw the tomato's possibilities, and now one can scarcely imagine southern Italian cooking without it.

20. One hemp strain is grown for the oil found in its seeds, another for its fibers (used for cordage), and a third for its leaves: marijuana.

21. The seeds you buy have gardening ease bred into them. In nature, they used to be just as picky as the seeds of other plants in nature, each one demanding its own special treatment, special shade or sun, special amount of water or dryness, special closeness to other plants or space to spread its leaves alone. The plants we call weeds have more of this adaptability than others.

22. The origin of every major staple crop plant—wheat, corn, rice, soybeans, and so forth—is lost to us. They all came about before the beginnings of written history. Because we have loved their beauty, we have brought some wildflowers under cultivation in relatively modern times. But we have not domesticated a major new food crop in at least two thousand years.

23. Sunflowers, the only world crop that was domesticated in what is now the United States, are grown commercially in Europe. One botanist, Edgar Anderson, points out how common this is: coffee originating in the Old World, but grown now mostly in the New; chocolate originating in the New World, and often grown now in the Old. "Rubber and quinine went to Malaysia from the New World," he says, and "bananas and citrus fruits came here from there." This is not so perverse as it seems. When a crop is grown for a long time, a whole constellation of pests also grow up to prey on it. "The farther you get from its center of origin the more of its pests can you hope to leave behind."

24. *Maíz reventador* is "the corn that explodes"—or popcorn.

25. Blueberries and cranberries, both members of the heath family, have always grown in New England. Both tolerate the acid soil of the region. Cranberries are grown in bogs.

26. New Jersey has five distinct regions: The well-known suburban-industrial corridor between Philadelphia and New York, a rocky region of farms and lakes in the northwest, a loamy area of farmland to the south, and the strip of seaside resorts culminating in the glitz of Atlantic City. The Pine Barrens is the fifth. Originally this gloomy forest covered almost two thousand square miles—a quarter of the state. Much of its outskirts have now been developed, but the empty center still holds, and it is situated precisely at the middle of the eastern megalopolis—an empty hole of nature in a doughnut of people and their works.

27. Bald cypress. They grow fast in the humid, murky canyons of San Antonio's central business district.

ANIMALS

Animals live and move; that's what distinguishes them superficially from minerals and vegetables, which, in the first instance, don't live or move, and in the second, live but don't move. More specifically, animals are living things that cannot produce their own food through photosynthesis, the way plants can. This gives them certain disadvantages, of course. But they have advantages too. There are about 1.3 million species of animal on earth, with many more still to be discovered, named, and studied, if we let them live long enough.

1. What replaced the very useful bison, or buffalo, when it was nearly exterminated on the North American prairies in the late 1800s?

2. Why are there polar bears in the Arctic and penguins in the Antarctic, but no penguins in the Arctic and no polar bears in the Antarctic?

3. How do polar bears hunt seals at the North Pole?

4. Fill in the blanks with animal types:
   Shetland _____
   Clydesdale _____
   Labrador _____
   Plymouth Rock _____
   Dalmation _____
   Komodo _____
   Polar _____

5. Where would you find the most venomous snake?

6. In what country was the world's first cat show held?

7. Does an endangered species always die out in the end?

8. In 1980 there were, in Zaire, about 370,000 individual specimens of a very large animal, the largest number in any African country. What is the animal?

9. Most primates are diurnal—they make their livings during the daytime. But one, the owl monkey or night monkey *(Aotus trivirgatus),* is nocturnal. On what continent might you find him?

10. Hawks and owls eat virtually the same food. How do they share the same habitats without seriously competing?

11. How high do birds fly?

12. Some camels have one hump and some have two. Where does each sort live?

13. Name the laziest animal in Africa.

14. On what continent would you find the tallest animal; what is its name?

15. What is the world's largest animal, and where would you stand the best chance of seeing one?

16. Go to a single cave in Kauchanaburi Province, Thailand. Enter. Look up. See the bats. They are Kitti's hognosed bats *(Craseonycteris thonglangyai),* and they have a wingspan of about six inches, wider than many birds. But they are actually the world's _____ (fill in the blank) mammal.

17. How big is the wingspan of the world's biggest bat:
Two and a half feet
Four feet
Six feet

18. In the Amazon Basin lives a creature that routinely grows large enough to straddle a sheet of paper without stepping on it with any of its eight legs. This is the largest _____ (fill in the blank) in the world.

19. What is the slowest land animal?

20. Where would you find the longest-lived mammal?

21. What country has a cockroach named after it?

22. There is a terrific rock band called the Del Crustaceans—they play mostly in Chicago. A crustacean is a creature of the crab ilk. How big can a non–Del crustacean get?

23. How long does it take the bats to exit Carlsbad Caverns in the evening?

24. What country has a rat named after it?

25. Twenty million bats congregate every summer in Bracken Cave in Texas. What do they all have in common?

26. What's the best way to control seagulls at airports?

27. The common starling *(Sturnus vulgaris)* has almost totally displaced the bluebird and the yellow-shafted flicker in North America. Where did it come from, and how did it do that?

28. Which of the dinosaurs ate humans?

29. Which was domesticated first, the dog or the cat?

30. Match these animal-named features with their site:
1 Dog Island   a Caribbean
2 Cat Island   Nicaragua,
3 Swan River   Honduras

4 Coyote Peaks
5 Fly River
6 Elephant Mountain
7 Kangaroo Island
8 Mosquito Coast
9 Pigeon River
10 Fish River
11 Snake River
12 Bay of Whales
13 Cape Dolphin
14 Deer Lodge
15 Goat Island

b Falkland Islands
c A town in Montana
d The island at Niagara Falls
e Antarctica's Ross Sea
f Idaho, Oregon, Washington
g Southwestern Namibia
h Northwest Minnesota
i Eastern Tennessee
j Western Australia
k Western Texas
l Western New Guinea
m Sierra Nevada Mountains
n Southeastern Bahamas
o Florida Gulf Coast
p South Australia

31. In 1983, Australia's tourism minister publicly described the koala as a "flea-ridden, piddling, stinking, scratching, rotten little thing." Lacking a defender in its own ministry of tourism, the koala (which is not a bear or even very much like a bear) found one elsewhere. Who stepped in to defend the koala?

32. What common trait do you find in animals that live in places where there is little or no light?

33. What is unusual about the droppings of desert animals?

34. How do certain desert tortoises cool off?

35. Match the horseracing track with its state:
1 Hialeah   a Maryland
2 Santa Anita   b Kentucky
3 Churchill Downs   c Florida
4 Pimlico   d California

37. Which is heavier:
a The average American woman
b The average grown-up female chimpanzee

38. What animal has been called "the swimming nose"?

39. The people of what country eat more chickens per capita than any other?

ANIMALS

1. Longhorns, Herefords, and Black Angus replaced the bison. Bison are much like cattle; if you had the skeletons of a bison and a Black Angus before you, you'd be hard pressed to tell the difference unless you looked at the horns. Bison are genus *Bison;* cattle are genus *Bos*. They can be interbred, however, and a modern reclassification would probably put them in the same genus. What we generally call "cattle"— from angus to zebu—would probably end up classified within the same species.

2. Geography is why. There's no reason why either couldn't survive in the other's place, but they have no way of getting to the place where they aren't. Both are cold-water creatures, by and large, and neither could survive a trip across the equator to get to the other pole. There is an equatorial penguin, but it couldn't live at the pole.

3. The polar bear hangs around beside places in the ice where a seal might stick its nose up to breathe. When it does, the bear smashes the seal's head against the ice with a paw.

4. Shetland pony, Clydesdale horse, Labrador retriever, Plymouth Rock hen, Dalmatian dog, Komodo dragon (a giant lizard), polar bear.

5. Sinuating near the Ashmore Reef in the Timor Sea off northwest Australia you'll find *Hydrophis belcheri*. It is six feet long, and the most poisonous of the sea snakes. We are told they are not aggressive. The venomousness of the Pacific sea snakes in general is one reason humans have not built a sea-level canal between the Atlantic and Pacific oceans. The locks of the Panama Canal keep the Pacific snakes out of the Atlantic.

6. The world's first cat show was held in London in 1871, despite a general Victorian dislike of felines. Cats have been welcome in Britain ever since. In certain suburbs of England, there may be fifty cats per acre.

7. The situation in Florida provides illuminating answers about species endangerment. The alligator, endangered twenty years ago, is now floundering across golf courses, sliding into backyard swimming pools, eating pet poodles, and menacing innocent retirees. Protection has given it a chance to return. That can be effective with the alligator, which lays hundreds of eggs and can coexist in the same territory as man.

   Things are more difficult for another Florida creature, the panther. Each Florida panther needs as many as four hundred square miles of wild country for hunting. Because settlement and development has destroyed so much habitat, there are only about twenty such panthers left alive.

8. The African elephant. Zaire's elephant population is rapidly shrinking due to poaching.

9. The night monkey lives in Peru and Paraguay, in South America, according to a Duke University biologist, Patricia Wright, who tracked him there. The monkey sees very well in low light, she found, and makes a low, hooting call much like that of an owl when searching for a mate. But only when the moon is full. They sleep during the day. Wright figures the night monkey stays up late to avoid daytime predators.

10. Hawks and owls may eat the same food, but they do it at different times. Hawks hunt in the daytime, owls

11. Some small migratory birds, in flocks, fly as high as 20,000 feet up, according to a research assistant at the American Museum of Natural History, Joe DiCostanzo. The bar-headed goose migrates from Asia to India and crosses over Mount Everest at an altitude of almost 30,000 feet. And one bird, a Ruppell's griffon—a vulture measuring 8 feet from wingtip to wingtip—flew into an airliner above the Ivory Coast in 1973. The plane's altimeter read 37,000 feet at the moment of impact. How can a bird breathe at altitudes where humans require a pressurized cabin? They have a system of air sacs, in addition to their lungs, which helps them to extract maximum oxygen from the air they breathe, even the thin, cold air over Everest.

12. The one-humped Arabian, or dromedary, camel lives in North Africa and the Mideast; the two-humped, or bactrian, camel, which also has longer hair, lives in central Asia.

13. The lion is probably the laziest. It sleeps at least twenty hours a day. However, it makes no sense to apply human words like *lazy* to an animal.

14. The African giraffe reaches 4.5 to 6 meters in height. It is still abundant in Africa's great game reserves.

15. The blue whale is easily the largest, at 190 tons or so in weight. Your best chance to see one would be on an extended Antarctic cruise during January (when it's summer there). Centuries of deep-sea whaling has reduced their number to something less than twenty thousand. They summer in polar waters and winter in warmer seas, where they breed.

16. Kitti's hognosed bats are the world's *smallest* mammal, if you judge them by weight. They weigh only seven-hundredths of an ounce.

17. The Bismarck flying fox (*Pteropus neohibernicus*) of the Bismarck Archipelago and nearby New Guinea has a wingspan of six feet and is the biggest damn bat you'll ever see.

18. Closely related representatives of three different genera, *Theraphosa*, *Lasiodora*, and *Grammostola*, are the largest spiders in the world. The sheet of paper we're talking about is the standard 8½-by-11-inch size, the size of this book.

19. The ai, star of crossword puzzles and Scrabble games, is also known as the three-toed sloth and slowest land mammal. You can find them in tropical forests of South and Central America, or more easily, in the city park of Puerto Limón, Costa Rica, where they hang upside down in the trees. The ai can be goaded

at night, thus occupying the same place but different ecological niches. (Incidentally, you can hear a hawk fly—its feathers whisper—but owls fly in eerie silence.)

to move through the trees at a top speed of fifteen feet per minute, though as the mat of moss and mold on its back suggests, it seemingly would prefer not to move at all.

20. The longest-lived mammal is found almost anyplace where mammals can live, and some where you wouldn't think it possible, because the answer is *Homo sapiens*. Humans live longer than any other mammal.

21. Germany. The German cockroach (*Blattella germanica*) is the most common of all the world's 3,500 cockroach species. Fifty-five species live in the United States, including the German. Okay, we may as well admit it; another cockroach has a country in its name, the American cockroach. A pair of cockroaches could theoretically produce 400,000 descendants in one year.

22. The biggest known crustacean, even bigger than Rick Telander, who plays lead guitar and sings a little for the Del Crustaceans, is the spider crab (*Macrocheira kaempferi*). It lives at a depth of 1,200 feet along Japan's Pacific coast and grows to be eleven feet across.

23. The bats of Carlsbad—about 3 million of them, of eleven different species—take as long as three hours to depart on their evening's hunt, during which they consume an estimated twenty-five tons of insects.

24. Norway has a rat named after it, even though it is a very tidy, pleasant country.

25. Every one of these bats is female, and they've come to give birth. No one knows why they do it here in Bracken's Cave. Or why their mates are still in Mexico, one thousand miles south, waiting for their return.

26. There's no best way to control gulls. Everything's been tried, from rubber snakes to shotguns to the conventional direct collision with airplanes—no method is perfect. At Pearson International Airport, in Toronto, trained falcons are on patrol from dawn to dusk.

27. The starling from Europe was introduced into Central Park, in Manhattan, in 1891. By 1940, these now-ubiquitous birds had spread to southern Florida, central Canada, and far California. Apparently, they displace other birds by competing more successfully for nest-holes.

28. No dinosaur ate people. Dinosaurs died out about 60 million years ago, and humans did not appear, even in rudimentary form, until 2 million years ago or so. But dinosaurs lasted for 100 million years and successfully fed on just what you'd expect. Some ate plants, some ate fish, others ate animals, including other dinosaurs, and some ate everything. The vegetarian dinosaurs ate mostly ferns, since they were the large plants around then. Mammals replaced dinosaurs; flowering plants replaced most ferns.

29. The dog was domesticated at least thirty thousand years ago, before any other plant or animal joined the human life-style. Some say cats were domesticated in Egypt five thousand years ago. Others say they haven't been yet.

30.

1–o	6–k	11–f
2–n	7–p	12–e
3–j	8–a	13–b
4–m	9–i and h	14–c
5–l	10–g	15–d

31. It was the Japanese who defended the koala. Six Japanese cities immediately wanted one for their zoos, and Tokyo set aside almost $2 million to purchase a pair, although it seems unlikely the Australians would ask that much for such an unpopular beast. A Japanese koala boom followed, in which all sorts of koala ties, toys, and T-shirts were marketed. Isn't that cute?

32. In places of little light, you find blind animals—blind fish, blind birds, blind crabs, blind eels, and assorted blind insects are known.

33. To conserve water, desert animals' droppings are extremely dry. Those of a camel, for instance, can be used almost immediately to fuel a fire. Some reptiles drop only "a cake of dry powder," in the words of one naturalist.

34. They simply pee all over their own hindquarters, and evaporation does the rest.

35.
1–c
2–d
3–b
4–a

37. You'll be pleased to know the average female chimp comes in at 75 pounds, much less than the average American female at 135 pounds. Female gorillas are another matter.

38. The shark. Some species can smell a small amount of blood in the water from a quarter-mile away.

39. The people of Saudi Arabia eat more chickens than any other. They—the chickens—are slaughtered in the Halal manner prescribed by Islam. The slaughterers dress from head to foot in gray cloth. They put on pointed hats. They face Mecca. They chant, "In the name of Allah." And they slit the chickens' throats. Saudi Arabia is an exporter of eggs.

II

THE HUMAN IMPRINT

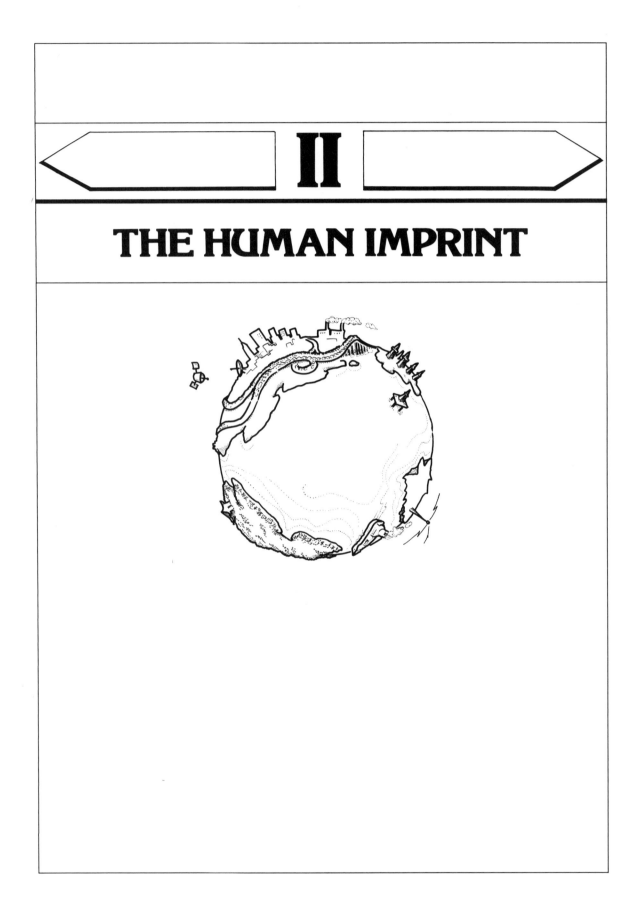

THE PEOPLED PLANET

Just as we regard the Earth as a special object in the celestial scheme of things, so do we regard ourselves as a special class of living creatures. Humans are animals, to be sure; but don't we also suspect we are something more?

The question is, Are we right?

The answer may not matter, however. We are here. We are definitely here.

Our sheer number is the most important fact about human beings. When Christ was born, the world held only 250 million of us. By the late Middle Ages, the number had doubled. That first doubling of our population took seventeen centuries. Today, despite the toll of two world wars; famines in the East, the Mideast, and Africa; the Soviet purges of millions of kulaks; and all the other disasters—natural and human-made—the human population has doubled in just the eight decades since the turn of the century. And the doubling under way now will take only thirty-five years. By 2010, we may number 7 billion.

What this proves is at least two things:

Humans as a group are extremely stupid.

And humans as individuals are determined to survive and prosper.

We may not be able to do both. Already the pressures of population have reduced the quality of life in many otherwise desirable places, from Sri Lanka to Italy to California. There is little doubt that this trend will continue unless we do something about it.

There is another salient fact about humans, besides our insanely growing number, and that is our incredible diversity, the complexity and beauty of our creations, the vast reach of our intelligence. The human career has been ever increasing our power to alter the planet.

When you combine the numbers with our evident need to dominate rather than harmonize with our environment and our equally evident lack of wisdom, what do you get?

Good question.

Here are some more:

1. What continent has the highest birthrate?

2. About how closely is any given living human related to any other?
   No more distantly than fiftieth cousins
   No more distantly than one-hundredth cousins
   No more distantly than five-hundredth cousins

3. According to Isaac Asimov, when will the earth's population reach its peak of growth, and at what number?

4. By and large, Amerindians and North Europeans are more compact—that is, have shorter limbs, smaller ears, and so on—than African Negroes. Why?

5. Think about the noses of your friends, who we assume represent many various human types. Why are they so different?

6. Who are the world's worst mathematicians?

7. Despite all our talk about ZPG (Zero Population Growth), only one country in the world had achieved it by 1983. Which one?
United States
Sweden
Canada
Kampuchea
Costa Rica
West Germany

8. Humans sweat to stay cool. Are there people—besides WASPs—who don't sweat?

9. Which country has the most lawyers?

10. What does steatopygous mean?

11. The average height of one Asian country's people has increased since 1945. What is the country and why have the people grown?

12. In 1487, the Aztecs of Tenochtitlán, situated where Mexico City is today, held a big ceremony to dedicate their Great Temple. It lasted from sunrise to sunset for four days. What did the ceremony consist of?

13. "Men bent on certainty," in Jared Diamond's phrase, have insisted on such oddities as chastity belts for their women. They have also employed other assurances of fidelity, such as female circumcision and infibulation (sewing the labia majora nearly shut). These are not merely ancient practices; they continue today, in about how many countries:
Three to five
Five to twenty
Twenty to thirty

14. What is an Aryan?

15. In what country are Japanese people Caucasians?

16. Political freedoms vary from group to group and country to country, but one group is almost universally denied full legal equality. One study concluded (fill in the blanks), "Nowhere are _____ fully equal to _____."

17. Match the country with its chief minority group:
1	United States	a	Swedes
2	Canada	b	Kazakhs
3	Belize	c	Ovimbundu
4	Suriname	d	Kurds
5	Argentina	e	Hungarians
6	Poland	f	Basques
7	USSR	g	Mayans
8	Romania	h	Gypsies
9	Hungary	i	Jews
10	Finland	j	Ukrainians
11	Spain	k	Hindus (South Asians)
12	New Zealand	l	Blacks
13	Mongolia	m	Maoris
14	Iraq	n	Quebecois
15	Angola		

18. The pill is one of the most efficient ways to reduce a nation's birth rate, though not always an easy one to persuade people to use. Among some peoples it is an article of faith that any attempt to foist the pill onto them is tantamount to attempted genocide. And researchers have found that in Iran, the pill goes down better when the husband gets it and gives it in turn to the wife—that way, he feels more in control. From such hints, and from what you suspect about various cultures and national characters, try to match the following countries with the percentage of women there who use the pill:
1	Hong Kong	a	2 percent
2	Thailand	b	8
3	Malaysia	c	24
4	Indonesia	d	11
5	Costa Rica	e	7
6	Egypt	f	12
7	Bangladesh	g	15
8	Honduras	h	16
9	Jamaica	i	3
10	Tunisia	j	22
11	Kenya	k	23
12	Turkey	l	25

19. Where, roughly speaking, are the world's four great concentrations of people?

20. The three countries with the highest rate of illegitimate births are all very close together on the globe. Name these three or the region they are found in.

21. What country has the highest population density per square mile of cultivated land?

22. Pick how fast the world's population grows:
a About 20 persons per minute
b About 150 persons per minute
c About 3,000 persons per minute

23. Where would you have found the most populous city in Jesus' day?

24. From which of these countries do the most people emigrate: Mexico, Iran, Brazil, China, East Germany? How come?

25. Only four large nations lost population between 1976 and 1981. Can you name any of them?

26. What country has the world's lowest birthrate?

27. Of the world's 4.7 billion people, about how many would you say are under arms?
  a 5 million
  b 26 million
  c 100 million
  d 600 million

28. Magellan is often termed the first man to sail around the world, and in a sense this is true—but not literally. Why not?

29. Why didn't American Indians invent and use the wheel the way Europeans did?

30. About how many people lived in North America when Europeans first arrived?

31. How many Native Americans remained in what is now the U.S. and Canada when de Tocqueville toured North America in 1831?

32. What European is usually said to have been the first to visit the site of New York City?
  Hudson
  Verrazano
  Champlain
  Raleigh
  Balboa

33. What Spaniard is credited with the discovery of the Grand Canyon?
  de Soto
  Coronado
  de Cárdenas
  Kino

34. Which European first navigated the Bering Strait?
  Bering
  Drake
  Cook
  Magellan

35. Who first landed in Antarctica?
  Scott
  Amundsen
  Byrd
  Kristensen

36. What exactly is "America"?

37. Where is the oldest ship in the world?

38. Match the game with its country of origin.
1	Badminton	**a**	Ireland
2	Basketball	**b**	England
3	Bowling	**c**	Uruguay
4	Handball	**d**	United States
5	Jai alai	**e**	North America
6	Lacrosse	**f**	Japan
7	Judo	**g**	British officers in India
8	Ping-Pong		
9	Water skiing	**h**	Basque Spain
10	Darts	**i**	Germany
11	Canasta	**j**	India

39. Fill in the final blank in this quotation from Paul Ehrlich: "No geological event in a billion years—not the emergence of mighty mountain ranges, nor the submergence of entire subcontinents, nor the occurrence of periodic glacial ages—has posed a threat to terrestrial life comparable to that of human _____."

40. Which is growing faster, the population of men or that of women?

41. According to an international study, the women of the world work about how many hours, as compared with the men of the world?
  One-tenth as many
  One-fourth as many
  One-half as many
  The same number of hours
  Twice as many hours

42. Where do women "enjoy sex eight times as much as men"?

43. What is a Moor?

THE PEOPLED PLANET

1. Africa has the highest birthrate, by a long shot. Some countries in Asia and South America are comparable to some African countries, but no continent comes close to this one. Most African women average six or more children. Kenya has the largest population-growth rate in the world—4 percent *per year*.

2. Geneticists are largely agreed that "no human . . . can be less closely related to any other human than approximately 50th cousin, and most of us . . . are a lot closer . . . ," geneticist Guy Murchie says. "The family trees of all of us, of whatever origin or trait, must meet and merge into one genetic tree of all humanity by the time they have spread into our ancestors for about 50 generations." The family of man, concludes Alex Shoumatoff in *The New Yorker*, "actually exists."

If you chart genealogy in a horizontal manner, for example (a recent vogue), you discover such curiosities as the fact that Jimmy Carter and Richard Nixon are sixth cousins (both descended from one Richard Morris, a New Jersey Quaker). If you chart it in the traditional way, vertically, you discover, as Murchie did, that "It is virtually certain . . . that you are a direct descendant of Muhammad and every fertile predecessor of his, including Krishna, Confucius, Abraham, Buddha, Caesar, Ishmael and Judas Iscariot. Of course you also must be descended from millions who have lived since Muhammad, inevitably including kings and criminals, but the earlier they lived the more surely you are their descendant." Shoumatoff writes hopefully, "If this vision of ourselves could somehow catch on, then

many of the differences that have polarized various subpopulations from the beginning of human history . . . would seem secondary."

3. Asimov believes that our population will peak at 7 billion—one and a half times the present figure—around the year 2010. Many of us will be here to see if he is right.

4. To reduce heat loss in the cold climates in which they have historically lived. However, we've all seen Scandinavians who are tall and skinny with long arms and big ears, as well as Africans who are squat, with small ears. These generalizations emerge only in statistical comparisons of many individuals.

5. Because of their parents' genes primarily, of course. Beyond that, nobody knows exactly. But the shape of an individual's nose correlates strongly with the climate in which that individual's ancestors lived, suggesting that natural selection for such functions as cooling, warming, and filtering have something to do with it: Consider the small noses of Orientals and Eskimos in a cold climate; the high bridges of Iranian and Amerindian noses in a dry climate; the broad and flat noses of some equatorial peoples in a moist, warm climate.

6. You can count on this: The world's worst mathematicians are the Nambiguara Indians of west-central Brazil. Their language contains no numbers.

7. Only Sweden had achieved ZPG by 1983.

8. Actually, white Anglo-Saxon Protestants do sweat—they just do it quietly and tastefully. Among people who actually perspire very little are the pygmies of Zaire—the Mbuti. Sweating does not cool the body very well in the humid regions of rain forest, where moisture just clings to the skin.

9. The United States has the most lawyers per capita—one for every 400 of us. That is three times the number in England, and twenty times that in Japan. We have 628,000 lawyers—twice as many as we had in 1960. Derek Bok, president of Harvard and a lawyer, says the U.S. legal system is "the most expensive and least efficient in the world." It soaks up forty thousand new lawyers every year, and still there is, Bok says, "far too much law for those who can afford it and far too little for those who cannot."

10. Steatopygous means having a fat ass. This condition is found most often in women, most often in bushmen and Hottentots in Africa, where each buttock can be as large as two feet across. Some think it stores fat for the bad season. Others think bushmen consider it attractive and the quality has been sexually selected.

11. The average Japanese man is about four inches taller today than at the end of World War II, largely because of the Westernization of the Japanese diet.

12. The ceremony consisted entirely of ritual killing. Without pause for four days, people were led four at a time to the altar and the obsidian knife. Sources vary, but it seems that between 10,600 and 80,400 people were killed.

13. The practices of infibulation and female circumcision are still common in at least twenty-three countries, from Africa through Arabia to Indonesia, according to Diamond.

14. An Aryan—from the Sanskrit word *Arya,* meaning "noble"—is a person who speaks an Indo-European language and who moved into northern India centuries ago. Today, a majority of the people of Sri Lanka are Buddhist Aryans. The term Aryan, although abused by various racists throughout history, Hitler being the most malignant, is properly a language-related term, not an ethnic one. "In our time millions of people fell before the Aryan myth of the Third Reich—a belief in a race of superior beings that was false to its very roots," as Peter Farb has written. "The Aryans were not a race but rather a great variety of peoples who spoke the early Indo-European (also sometimes known as Indo-Aryan) languages. Rather than being primeval Germans, they were Asiatic invaders who settled the Iranian plateau and the Indian subcontinent, from which centers their language spread westward to the Near East and Europe."

15. Japanese are considered honorary Caucasians in South Africa—otherwise that country's race laws would prohibit the Japanese from free movement and thus from providing all sorts of goods and services

that the South Africans want. Japanese were declared "white," whereas local Chinese remained "coloured."

16. The full statement is "Nowhere are women fully equal to men."

17.
1–l (12 percent)	9–h (6 percent)
2–n (28 percent)	10–a (7 percent)
3–g (15 percent)	11–f (6 percent)
4–k (15 percent)	12–m (8 percent)
5–i (1 percent)	13–b (5 percent)
6–j (16 percent)	14–d (16 percent)
7–j (3 percent)	15–c (35 percent)
8–e (8 percent)	

18.
1–k	7–i
2–j	8–d
3–h	9–c
4–g	10–e
5–l	11–a
6–f	12–b

19. The earth's four population concentrations are in eastern North America, in western Europe, in southern Asia (India, Bangladesh), and in eastern China and Japan.

20. Sixty percent of the births in Panama, Guatemala, and El Salvador are to unwed mothers. These are all countries of Central America. The fourth-ranked is the Dominican Republic. One guess as to why these rank so high in illegitimacy is that they are undeveloped countries located very close to the developed world—by which they have been carefully scrutinized, counted, and judged. In matters of this sort, a lot depends on how you define "marriage," thus, on who does the defining.

21. Japan, with 6,153 persons per square mile of cultivated land, on the average. In Hong Kong, the British Crown Colony that abuts the People's Republic of China, from which many flee, there is a single square mile on which more than a *half million* people live. Most of Hong Kong's food supplies come from China.

22. Our numbers grow by about 144 humans every minute, so the answer is b. Gives you thought for food, doesn't it?

23. Seleukia, near Baghdad, had something over a half million persons at the time of Christ. The cities of the Middle East were the most populous in the world, generation after generation, from as long ago as we can measure, until about this time in history, when the West began to grow. By the early 1800s London was the most populous city. Nowadays, that distinction belongs to either Tokyo or Mexico City, depending upon whose figures you believe and where you place the city limits.

24. Mexico loses more of its citizens each year than any

other country. In 1976, for instance, 800,000 came into the United States from there—many of them illegally. There is a tempting contrast in standard of living between a populous developing country like Mexico and a developed one like the United States. And a border 1,933 miles long is not easy to blockade. More people immigrate to the United States annually than into any other country—and not, of course, only Mexicans.

25. These four nations lost population recently: the United Kingdom (because of emigration and a falling birthrate); East Germany (a low birthrate); Lebanon (emigration and war); Kampuchea, formerly Cambodia, (war, slaughter, and emigration of survivors).

26. The Vatican has the lowest birthrate. It is also the smallest country in the world at about 109 acres, and at last count the least populous—with only 728 inhabitants.

27. There are about 26 million humans under arms these days.

28. Magellan's expedition did manage to circumnavigate the globe, under the leadership of Captain del Cano, but Magellan didn't. He was killed partway there, in the Philippines.

29. They did invent the wheel. They used it for pull toys—little dogs and caimans on wheels for kids. But they didn't bother to make wheeled carts for real work, perhaps because they had no draft animals to pull them across their rough terrain (the horse had once existed here but had become extinct; the Europeans brought it back, which radically altered the culture again). Also, they had slaves. It was easier to let the slaves carry things.

30. About 10 to 12 million people lived between the Rio Grande and the Arctic Circle.

31. Only about 250,000 Native Americans remained by the time de Tocqueville came. The rest had been killed off by disease, starvation, and whites.

32. Verrazano, in 1524.

33. de Cárdenas, in 1540.

34. Bering, in 1728.

35. Kristensen, in 1895.

36. America is all of the New World, not just where people of the United States of America live, which is why it annoys a Uruguayian or a Cuban to hear the U.S.A. called "America," as if it were the only

America. Canadians in particular dislike this presumption.

37. The oldest surviving ship is Cheops's boat, which was carefully dismantled 4,600 years ago and lay in 1,224 pieces in a pit alongside the Great Pyramid until it was discovered in 1954. It has been rebuilt and now can be seen in a museum there.

38.

1–j	7–f
2–d	8–g
3–i	9–d
4–a	10–b
5–h	11–c
6–e	

39. *Overpopulation* is the word that completes the Ehrlich quotation.

40. There are more women on earth than men already, and by the year 2100 they will outnumber men by nearly 175 million. The number of women had risen to 2.41 billion by 1985, according to a study by the U.N. Fund for Population Activities. By the end of the century, there will be 3 billion or so. One thing this means, the report found, is that female illiteracy is on the increase, because of poor education for women combined with the rapid growth of the female population. In 1980, 60 percent of the 824 million illiterates in the world were women.

41. Women of the world work about twice as many hours as men. Most of it is unpaid labor.

42. Evidently, it is in India where women enjoy sex eight times as much as men; the quotation is from the ancient Indian text, *Mahabharata*, which is far from the only erotic ancient Indian text still widely respected and consulted. The great hero of this tale is Krishna, Vishnu's eighth incarnation, a handsome youth of blue hue. As we men know, you can never tell what a woman will admire.

43. A Moor is a Moslem of mixed Berber and Arab descent. It was these people who invaded Spain in the eighth century and occupied it until 1492. The effects of this period of occupation, according to many scholars, have lasted into the present day because they've contributed to the formation of the Spanish, and Latin American, characters. Some scholars attribute the Latin American taste for government by strong-man to Spain's occupation by the Moors and the subsequent consolidation of power by Spanish kings.

FOOD

We won't say you are what you eat, because it's been said too many times and isn't true. Food is crucial in itself, as fuel; but you are flesh and fluids and mind. No matter what you eat, the biochemistry of the digestive system and the reliability of genetic coding mean you'll end up looking and acting pretty much like the rest of us.

The more complex we humans become, the more our food reveals about us. Primitive people eating grubs and berries aren't making a statement—grubs and berries may be all there is to eat. Or they taste better than the other choices. But just look at your friends: the vegetarian, the meat-and-potatoes person, the feeder-on-fast-foods, the health-foodist, the sugar buff, the boozer, the chocolate freak, the pasta pest, the picker-at-plates, the gourmet. Are they merely eating? Are they also inventing, creating? Are they making a statement— taking a stand of some kind with their tofu and sushi, their pasta *al pesto* and Szechuan chicken with cashews, their burgers and buns and bagels and lox and garlicky sauces and Häagen Dazs?

Maybe they're just hungry. Maybe they are also saying that they're open to new experience, or that they're loyal to their cultural roots, that they love self-indulgence, or that they fear the novel, or that they want out? And that's just your friends. Think what must go on across town, across the country, across the world.

"Read any month's itemized grocery bill for a thrilling chapter of applied geography," the *National Geographic* said in 1942, in a story on the world's food.

It could as easily have been about the food of America. Cuisines from all over the world have been brought to this country and embraced happily by enough Americans to maintain something of their ethnic originality. Many Americans, of course, remain boring eaters—you see them packing the McDonald's in Paris and the Kentucky Fried Chicken stand in Nairobi.

Evolutionarily, we humans are omnivores. We developed by eating whatever would nourish us, and vegetarianism, however popular it seems to be, appeared only a few thousand years ago and is rather a luxury. It takes not only some leisure, but intelligence and attention to the details of protein management to be a successful vegetarian.

1. If you order "mountain chicken" in Montserrat or Dominica, what do you get?

2. On an ordinary sort of warmish day, how much of the food you eat goes to maintain your body temperature?

3. What color is the filling of proper key lime pie?

4. About how many pounds of food does the average American eat in a year?
500 pounds
1,500 pounds
5,000 pounds
8,500 pounds

5. What U.S. state is famous for its potatoes?

6. Who was responsible for the Irish Potato Famine:
   a Pizarro
   b Rhodes Murphy
   c Verrazano
   d Cézanne

7. Why did Ireland's population more than double, to 8 million souls, between 1754 and 1845?

8. Has anyone ever invented an animal?

9. The average human body needs 2,500 calories a day to maintain itself. What proportion of humanity gets this many?

10. Among the four major staple crops—potatoes, wheat, maize (corn), and rice—which is the oldest?

11. Name the home country or region of the following dishes, and describe the dish:
    1 Panettone   10 Cioppino
    2 Kielbasa   11 Bombay duck
    3 Gazpacho   12 Flageolets
    4 Chapati   13 Spaetzle
    5 Sopaipillas   14 Rehrücken
    6 Kimchi   15 Couscous
    7 Havarti   16 Baklava
    8 Piroshki   17 Drunken clams
    9 Borscht   18 Chitterlings

12. Cooks in what country or region first used each of the following spices or herbs? Match herb or spice with source region on right.
    1 Basil   a Turkey
    2 Bay leaves   b Southern Russia
    3 Chives   c Southeast Asia
    4 Fennel   d India and China
    5 Tarragon   e Tropical America
    6 Allspice   f India
    7 Capsicum   g Nile Valley
      (chili, paprika)   h China
    8 Caraway   i Caribbean
    9 Cinnamon   j Mediterranean
    10 Clove   k South and Southeast
    11 Cumin   Asia
    12 Ginger
    13 Vanilla

13. By variously combining the twenty-four spices most commonly found on any grocery's shelves, you could create precisely 17,777,215 different flavor combinations. Name the four most popular spices in the United States.

14. California produces a great many oranges. Does anyplace in the United States produce more?

15. What city's name means "house of bread"?

16. Which U.S. president was at once so "American" that he was the first to combine French fries with steak, so cosmopolitan that he was the first to serve spaghetti at a state dinner, and though moderate in his drinking habits, sport enough to spend two thousand dollars a year on imported wines and die forty thousand dollars in debt?

17. Name the country of origin of the following dishes:
    Swiss steak
    Chop suey
    Cioppino
    Russian dressing
    Vichyssoise
    Hamburger in a bun

18. Where would you find the McDonald's that has served the most hamburgers in a single day?

19. What happens to the millions of cows that inevitably die in India, a largely Hindu and vegetarian nation in which cattle are sacred?

20. About how much meat would a group of one hundred New Guinea cannibals consume?

21. This creature, the (name the animal), though cooked and eaten for its protein in many nations, is permitted by the natives to roam freely in (name the country), where it is cared for and treated with veneration.

22. What is the most widely used flavoring agent in the Third World?

23. Here are some characteristic flavor combinations; match each with its country or region:
    1 Tomato, cumin, chilis   a Russia
    2 Soy, ginger, garlic   b France
    3 Garlic, cumin, ginger,   c Southern Italy,
      turmeric, coriander,   southern France
      cardamom, pepper,   d Greece
      mustard seed, and   e Mideast
      so on   f Iran
    4 Yogurt, dill, mint   g India
    5 Lemon, parsley, garlic   h China
    6 Lemon, oregano   i Mexico
    7 Olive oil, tomato,
      basil, oregano, garlic

8 Butter, wine, cream,
stock

9 Sour cream, dill,
caraway

24. Where is the World Kielbasa Festival held?

25. A one-part chicken question: Where do the small chickens known as bantams come from?

26. A two-part chicken question:
a How many chickens do Americans eat in a year?

b How many chickens would it take to provide every human on earth with a chicken?

27. What is "the first human food to be proved epidemiologically as a cause of human cancer"?
Human brains
Salted fish
Southern fried chicken
Charbroiled steak

28. A final chicken question: If you order "bamboo chicken" in Belize, what do you get?

FOOD

1. Frogs' legs the size of chickens.

2. On an ordinary day, we use half our food intake just to stay at 98.6 degrees F. When it's cold, that amount rises.

3. The filling of real key lime pie is yellow, because key limes are yellow when ripe. Persian limes are green, and the only key lime pie they make is fake.

4. The average American consumes about 1,425 pounds of food per year, according to the U.S. Department of Agriculture. We're eating more healthful food than in the past, too, they say. We are buying far more citrus, apples, avocadoes, grapes, plums, yogurt, broccoli, corn, cheese, chicken, cauliflower, and lettuce than ever before. We're using more low-fat milk. We're eating less lard, lamb, beef, and butter, and drinking less whole milk. On the other hand, we're eating fewer potatoes, which are good for us and eating a lot more sugar, which is rotten for us.

5. Idaho. But then people from Maine, North Dakota, Oregon, and several other states would probably argue this important point.

6. The Potato Famine of the 1840s could not have happened without the explorations of Pizarro, the conquerer of Peru. Potatoes were first grown in the Andes and came to Ireland only in the seventeenth century.

7. Because of the introduction of the potato, which the Irish almost seemed to be waiting for. All at once, a single acre could feed five or six people and their animals for a year.

8. The domestic breeds of dog were all invented, more or less, over centuries by breeders seeking certain characteristics—cute eyelashes or long teeth and nasty dispositions, and so forth. Mules are reinvented, in a sense, every time a farmer crosses a horse with a donkey. And Jacques Makowsky, a retired designer of perfume packages, invented an entirely new creature, the Rock Cornish game hen. It launched him on a second career, selling the little, single-serving birds to "gourmet" restaurants and later to supermarkets. Genetic engineering promises even more extraordinary "inventions."

9. About half of us get enough to eat. The rest of the people on earth are hungry, malnourished, or starving.

10. Wheat seems to be the oldest of the staple crops. Cultivation probably began more than ten thousand years ago.

11.
   1 A traditional Italian fruitcake
   2 Polish sausage
   3 Spicy Spanish tomato soup, served cold with chopped vegetable garnishes
   4 An unleavened fried bread from India
   5 A puffed, fried bread from Mexico
   6 Korea's answer to sauerkraut—a spicy fermented cabbage
   7 Danish cheese made from skim milk
   8 Little meat turnovers from Russia, known throughout Slavic countries
   9 Beet soup from the Slavic countries
   10 Seafood soup, usually with shrimp, clams, lobster, and several fish varieties—from Monterey County, California, U.S.A.
   11 This Indian Ocean fish is salt-dried and then curried
   12 Small green kidney beans from France

13 Tiny central European dumplings, terrific with Hungarian goulash

14 Viennese almond cake molded to resemble the back of a deer

15 Chicken or mutton stew from North Africa

16 An eastern Mediterranean dessert of pastry leaves and honey

17 Clams in chicken broth with ginger, from eastern China

18 Fried pig intestines from the United States: say "chitlins"

12.

1–f	8–a
2–j	9–c
3–h	10–c
4–j	11–g
5–b	12–d
6–i	13–e
7–e	

13. The four most popular spices in the United States, according to the American Spice Trade Association, are black pepper, cinnamon, chili powder, and paprika. They are followed by oregano (which is properly an herb, not a spice) and cloves. Other popular ones are sage, nutmeg, mustard, poultry seasoning, and ginger. There are also what are called in the trade "convenience spices": garlic powder, garlic salt, onion flakes, and parsley flakes.

14. Polk County, Florida, alone produces more oranges than does California.

15. Bethlehem means "house of bread" in Hebrew. There are Bethlehems in Pennsylvania, Israeli-occupied Jordan, and the Orange Free State, South Africa.

16. Thomas Jefferson.

17. All these dishes were invented in the United States.

18. The McDonald's that has served the most burgers in a day is in Enoshima, Japan. On each New Year's Day in recent years, this outlet has served more than seventeen thousand hamburgers.

19. Most of the dead cows in India are eaten, although not much is said about it.

20. A typical group of one hundred cannibals would consume about a person a week.

21. The venerated creature is the dog, and the countries are many, including our own. Dogs are served—as guests—in many French restaurants.

22. The chili pepper is the most widespread flavoring agent; a quarter of adult humans use it. Its basis is capsicum—the hot stuff—which makes you sweat, thus lowering body temperature through evaporation. It also stimulates the appetite and contains vitamins A, B, and C. But why do so many people eat something that is so unpleasant at first? Nobody likes chili the first time, says Dr. Paul Rozin, a psychologist at the University of Pennsylvania. And the mouth never becomes desensitized, even if you eat the stuff for years. Rozin thinks our motive is "benign masochism"—that is, thrill seeking. We also learn to love cold showers and roller-coaster rides, he says. Maybe we get a kick out of breaking some rules, once we learn it's harmless.

23.

1–i	6–d
2–h	7–c
3–g	8–b
4–f	9–a
5–e	

24. Chicopee, Massachusetts, holds the Keilbasa Festival. You'd think it would be in Detroit, Chicago, or Milwaukee.

25. Bantams come from eggs, of course, via the Bantam region of Java, in Indonesia.

26. **a** over four billion chickens.
b over four billion chickens.

27. The first human food definitely linked to cancer, according to Brian Henderson of the University of Southern California Comprehensive Cancer Center, is a salted, partially rotted fish commonly eaten by Chinese from early childhood. It causes a silver-dollar-sized cancer at the back of the throat—naso-pharyngeal carcinoma—which is the leading cancer among Chinese men and ranks third or fourth among women. The researcher believes the cause may be a chemical formed during the rotting process. This fish dish is used as a weaning food among the Chinese, and the cancer in question shows up fairly early, between ages fifteen and twenty-five.

28. "Bamboo chicken" is roast iguana.

DRINK

Drink is odd. Even the word is ambiguous—when you say "drink," you may mean water or any other liquid, or you may mean, specifically, alcoholic fluids. If you say, "Does he drink?" you may mean "Does he drink alcoholic drinks at all?" or "Does he drink alcoholic drinks too much?" but almost never "Does he imbibe any fluids?"

There's a long way to go in understanding how and why the patterns of alcohol consumption differ among various peoples. They have to do with religion (Muslims are not permitted to drink) and with money (making

liquor is a good way of turning your bulky, perishable corn crop into cash), and with culture (the Jews tend to drink less than the Irish), and perhaps with physiology (some people become alcoholics easier than others, and it is possible that some peoples, such as the Amerindians, have a physiological intolerance for alcohol).

One thing is certain: Alcoholic drink is found in virtually every culture, including those in which it is ostensibly banned.

1. For openers, what's the name of the beer in Tombouctu?

Bamako	Tim Buck II
Tombigbee	Bou Bou Tou
Buckhorn	

2. Match these beer brands with their country of origin:

1	Tusker	a	Mexico
2	Foster's	b	Netherlands
3	Carta Blanca	c	Japan
4	San Miguel	d	Namibia
5	Molson	e	U.S.A.
6	Coors	f	West Germany
7	Swakop	g	Philippines
8	Heineken	h	Kenya
9	Beck's	i	Australia
10	Asahi	j	Canada
11	Primus	k	Zaire

3. Match the liqueur with its country of origin:

1	Grand Marnier	a	Italy
2	Nassau Royale	b	France
3	Cointreau	c	Netherlands
4	Kahlua	d	Mexico
5	Praline	e	New Orleans
6	Vandermint	f	Bahamas
7	Drambuie	g	Israel
8	Chambord	h	South Africa
9	Van Der Hum	i	Scotland
10	Sabra		
11	Chartreuse		
12	Amaretto (original)		
13	Frangelico		

4. Where does real sherry come from?

5. In what country do they drink ouzo?

6. If the tannin in tea tends to cause esophageal cancer, as Dr. Julia Morton, the botanist, and others believe, and if the Japanese and the English are both tea-loving peoples, as they are, why is the esophageal cancer rate high in Japan and low in England?

7. How many gallons of beer are in a U.S. barrel?

8. How much bubbly (in milliliters) does the traditional champagne bottle hold?

9. Champagne comes in other sizes, too, with aptly evocative names. Arrange these in ascending order of size: balthazar, magnum, nebuchadnezzar, salmanazar, methuselah, jeroboam, rehoboam.

10. Where did coffee originate?

11. Who drinks the most coffee?

12. Where did mocha java get its first name?

13. Match these world regions with the development of these varietal wines:

1 Bordeaux	a Pinot Noir
2 Cape of Good Hope	b Colombar
3 Loire Valley	c Riesling
4 Burgundy	d Cabernet Sauvignon
5 Lyon	e Syrah
6 Avignon	f Gamay
7 Midi	g Furmint
8 Tuscany	h Semillon
9 Rhine Valley	i Chenin Blanc
10 Sauternes	j Pinotage
11 Cognac	k Carignan
12 Hungary	l Sangiovese

14. What is the "Caledonian flu"?

15. What kind of beer is Tsingtao?

16. Where does Calpis come from?

17. The former prime minister of what country held a Guinness world record for the greatest amount of what beverage consumed in the shortest time?

18. What is single-malt Scotch?

19. What kind of girl is a Saint Pauli girl?

DRINK

1. Bamako is the beer of Tombouctu.

2.
1–h	7–d
2–i	8–b
3–a	9–f
4–g	10–c
5–j	11–k
6–e	

3.
1–b	8–b
2–f	9–h
3–b	10–g
4–d	11–b
5–e	12–a
6–c	13–a
7–i	

4. Sherry comes from Spain, from the town of Jerez. If you slur Jerez just right, you discover where the English word *sherry* came from.

5. Greece.

6. Because the English add milk to their tea, which binds the tannin and prevents its absorption. This whole question of tannins and esophageal cancer is an interesting one, and full of surprises. The people of western Kenya, who make their beer and mash from a high-tannin sorghum plant, have a cancer rate fourteen times higher than that of the neighboring tribe, who make their beer from honey. The cancer was rife among the Dutch one hundred years ago when they were a tea-drinking (sans milk) people, but after they developed coffee plantations in Java and changed their national beverage to coffee, the cancer rate dropped. In the United States, the esophageal cancer rate is relatively low, except among people who drink a lot of tannin-rich herb decoctions—and also among black men, where the rate has been rising alarmingly. Morton attributes this to the high levels of tannin in cheap red wine.

7. There are thirty-six gallons of beer in a U.S. barrel.

8. There are 750 milliliters in the traditional bottle of Champagne. This is the same size as the usual European wine bottle and slightly smaller than the traditional four-fifths quart used until recently in the United States.

9. The magnum (at 1,600 ml) is next larger from the standard bottle, followed by jeroboam (3,200), rehoboam (4,800), methuselah (6,400), salmanazar (9,600), balthazar (12,800), and the immensely thirst-quenching nebuchadnezzar which holds 16,000 milliliters, or more than four U.S. gallons.

10. Coffee came originally from Ethiopia, where coffee trees still grow wild in the mountains. It spread with Islam in the seventh to tenth centuries.

11. The people who drink the most coffee are not the Brazilians, Colombians, Peruvians, Costa Ricans, or even the North Americans, but the Finns, whose consumption is twenty-eight pounds per person per year. It is, of course, cold in Finland, and Finns, living as they do on the Soviet border, like to stay alert. Brazilians drink about eight pounds per year, and we Americans drink twelve.

12. Mocha was the former coffee capital of Yemen.

13.
1–d	5–f	9–c
2–j	6–e	10–h
3–i	7–k	11–b
4–a	8–l	12–g

14. A hangover earned by drinking single-malt Scotch whisky.

15. As is obvious from the name, which means "green island," Tsingtao is Chinese beer. But in a more real sense it is German, because the Tsingtao brewery was set up over one hundred years ago by German brewers who leased property from the Chinese government for the purpose, and who brought with them a German beer formula still used by the Chinese brewers who now operate the place. As might be expected, Tsingtao is wonderful beer: deep and strong and full-flavored. Many Chinese restaurants in the United States serve it too cold.

16. From Cals, of course! Ha ha, that's a common occidental's joke in Japan, having to do with a brand of lactic bacterium drink called Calpis. The Japanese enjoy a good lactic bacterium drink, whatever that is.

17. Bob Hawke, a prime minister of Australia, once held the world record for most beer consumed in the shortest time. Australia is the kind of place where such a record helps more than it hurts.

18. Single-malt Scotch whisky is the unblended Scotch made and bottled by a single distiller. It is analagous to a château-bottled wine.

19. A Saint Pauli girl is presumably a hooker, since the Saint Pauli neighborhood of Hamburg, Germany, was historically a red-light district.

SHELTER

Maybe we like to go camping to prove to ourselves that we can still survive without all that superfluous shelter we—the haves, that is—have erected around ourselves. We really don't need insulation, wall-to-wall carpeting, track lights, and Levelors. At least not to keep the weather out. But it is true that most of us cannot live in comfort without some shelter from weather, dangerous animals and people, and bugs with too many legs.

So we build shelters. We always have. Sticks and stones may break our bones, but piled and connected, they also have the power to save us. A ring of rocks at Olduvai Gorge, circa 1.7 million years ago, is probably the remains of a very early shelter. People were certainly building shelters in northern Europe 500,000 years ago.

1. What is the best housing deal in the United States?

2. Define the following:
   a Yurt
   b Hogan
   c Tepee
   d Wigwam
   e Shoji
   f Tatami
   g Wattle and daub
   h Adobe
   i Soddy
   j Chikee

3. What are all the following? Name the city and country in which each is found.

Shepherds	Fairmont
Mandarin	Fountainebleau
Kahala	Ritz
Chelsea	Hassler
George V	Claridge's
Algonquin	

4. What and where is Drop City? What are the shelters there made from?

5. Say a little about the etymology of the word *gaudy*, if you can.

6. Where does the bungalow come from?

7. Where does the log cabin come from—and where might you find the oldest one in the world?

8. What is a *tokonoma?*

9. There's a great Moorish palace in Spain, made of red brick. The color of the brick is a hint to its name, which is: _____

10. In the center of the typical Latin American town you will invariably find the same three things. Name two of them. Or three if you can.

11. Until recently a city-dwelling Chinese was allotted

about 3.6 square yards—about the area of a generous dining-room table—for a particular use. What was the use?

12. Match these castles and palaces with their country:

1	Morro Castle	**a**	China
2	Windsor Castle	**b**	U.S.A.
3	Hermitage	**c**	France
4	Casa Loma	**d**	Cuba
5	Versailles	**e**	USSR
6	Iolani Palace	**f**	England
7	Forbidden City	**g**	Canada

13. In which Latin American country have many Japanese immigrants found both shelter and prosperity?

14. What does *defenestration* mean?

15. Just about all humans once lived in caves, didn't they?

16. Who said, "The house is a machine to live in"?

17. Who said, "Form follows function"?

18. Who said, "Less is more"?

19. In this quotation from Black Elk, a member of the Oglala Sioux, fill in the blank, and translate the word *Wasichus:* "The life of a man is a circle from childhood, and so it is in everything where power moves. Our _____ were round like the nests of birds, and these were always set in a circle, the nation's hoop, a nest of many nests, where the Great Spirit meant for us to hatch our children. But the Wasichus have put us in these square boxes. Our power is gone and we are dying, for the power is not in us anymore."

20. No matter where one builds a shelter, obviously it has a shape. Frank Lloyd Wright had a theory about the meaning of various geometric shapes used in shelters. Match the shape with what it signified to Wright:

Circle   Integrity
Square   Infinity
Triangle   Aspiration

21. Can you name a material that was very common in ancient Egypt for the construction of houses but is no longer used? Hint: It had another, familiar, commonplace use.

22. In what major way did the Parthenon look different to the ancient Greeks than it does to us?

23. Lords built castles during medieval times. About how many were built in Germany, for example, where a favorite tourist excursion is the drive or boat trip along the Rhine?
Five hundred
One thousand
Ten thousand

24. When English settlers came to North America, they naturally built houses the way they had back home—framed with big timbers and the spaces filled in between with mortar. You've seen a few of these so-called half-timbered houses. But not many. Why?

25. Can you name a type of house that was born in the United States?

26. Walk around the unreconstructed older sections of certain European cities and see the weathered old gray buildings of the Middle Ages. What was different about them then?

SHELTER

1. Build-it-yourself. About 150,000 American families build their own house every year—either subcontracting the work, or doing it all themselves. Not only does an owner-built house cost less than one built by a developer, but the workmanship will almost certainly be superior (no matter how inexperienced you are in the beginning, you will learn fast and care more, and it doesn't take much brains to learn what carpenters and plumbers do). And if you are wise and disciplined you could pay-as-you-go and end up, like 40 percent of those 150,000 families, with no mortgage payments ever. Think how that would raise your standard of living.

2. **a** A portable, circular shelter made of sticks, canvas, and felt, carried on two camels or horses, used by various ethnic groups from Iran to Mongolia for thousands of years.
   b House of poles or logs, plastered with mud, Navaho.
   c Plains Indians tent of skins.
   d A rounded lodge of poles with bark or skin covering.
   e Rice-paper screens in a Japanese house.
   f Floor mats in a Japanese house. They're of standard size (about three feet by six feet) and the area of a Japanese house is determined by the number of mats it will have, rather than by square footage.
   g A building method, using mud plastered over woven branches.
   h Construction of dried mud blocks.
   i Early Nebraskan sod house, 1860–1900.

   j Miccosoukee thatched platform on hammock in Everglades.

3. These are hotels. Shepherds is in Cairo, the Mandarin is in Hong Kong, the Kahala Hilton is in Honolulu, the Chelsea is in New York, the George V is in Paris, the Algonquin is in New York, the Fairmont is in San Francisco, the Fountainebleau is in Miami Beach, the Ritz is in Paris, the Hassler Plaza is in Rome, and Claridge's is in London.

4. Drop City—now a ghost town—was an acid-vision, hippie dome commune near Trinidad, Colorado, subject of a book by one Peter Rabbit. The shelters were domes made from car tops, with windshields for fenestration.

5. The word *gaudy* comes from the name of a man, a Spanish architect named Gaudi, of Barcelona, who designed buildings, including cathedrals, of flamboyant shape.

6. The bungalow is from India (*bangla*—of Bengal). It was adopted by the English in the mid-nineteenth century, then adapted in the United States in the early 1900s, especially in California and Florida. In Tampa alone, four thousand bungalows were built between 1910 and 1930. What is a bungalow? A low, one-story house, with a colonnaded veranda and a pitched roof with deep eaves. They tended to be small, inexpensive, and cozy, with a porch and an open plan inside (rather than boxy little rooms) that clearly influenced contemporary architects and the way we live today.

7. The log cabin was brought to North America by German and Scandinavian immigrants during the late

seventeenth century. They made ideal homes for pioneers, since they were easy to erect from local materials. For the oldest in the world, you might look in Norway. One, in the Oslo Folk Museum, dates from A.D. 1250.

8. In traditional Japanese houses, the *tokonoma* is an alcove in which a decorative object such as a picture, bonsai tree, or flower arrangement is displayed. The display is changed frequently.

9. The palace is the Alhambra, built in the thirteenth century, with lavish decoration added a century later. *Alhambra* means "red" in Arabic.

10. The center of most every Latin American town has a plaza with a church at one end and government offices along at least some of the other sides.

11. Until recently, the average city-dwelling Chinese was allotted the 3.6 square yards of space to make his home in. Population control and an ambitious building program have permitted allocation of more space.

12.
1–d   5–c
2–f   6–b
3–e   7–a
4–g

13. Japanese-Brazilians comprise ½ percent of Brazil's population. However, their farms provide Brazil with 60 percent of its soy beans, 71 percent of its potatoes, and 94 percent of its tea crop.

14. To "defenestrate" something is to throw it out the window. *Fenestration* is the name given to windows and window treatments in architecture. Defenestration used to be a popular way of getting rid of royalty in central Europe.

15. No, they didn't. Not many of those early people actually lived in caves, not because caves aren't sometimes good places to live, but because there aren't that many caves, and even back then, there were too many humans. The reason you find so much evidence of human habitation in caves is that caves—unlike open spaces—preserve that evidence.

16. Le Corbusier.

17. Louis Sullivan.

18. Mies van der Rohe.

19. The word in the blank should be *tepees*, and the word *Wasichus* means "whites," or "Americans."

20. To Wright, the circle signified infinity (as it seemed to, interestingly, to Black Elk), the square integrity, the triangle aspiration. You can see this in his work; the triangle, for instance, being used in churches where it not only mimicked the shape of praying hands but expressed human aspirations to something higher.

21. The Egyptians built many houses from papyrus, the familiar fibrous reed they also used to make paper, as well as boats, sails, and cloth.

22. The Parthenon is beautiful to us for its line and proportion, and perhaps even for the lovely, quiet tastefulness of its plain marble. But it looked much different to those who built it. For one thing, all the friezes and other sculpted areas above the columns were painted in bright red, blue, and gold.

23. Germans built at least ten thousand castles in the medieval period.

24. You don't see many half-timbered English-style houses in the United States because not many were built. Cute as we may think they are, they were terribly ill-suited to the climate in New England. The wood shrank; the mortar dried, cracked, and fell out, allowing icy gales to whip through living rooms. New England towns soon looked different from English towns, because the houses were covered almost immediately with clapboards, invented to keep the weather out.

25. Many kinds of shelters were created in what is now the United States—the tepee, for example. But the adobe house is one of few actual houses born here. Made of whitewashed, sun-dried brick, with a nearly flat roof of tiles, it stays cool in hot summers and is easy to warm in mild winters. Thus, the adobe is perfectly suited to the climate of the Southwest, where it was invented. In Santa Fe today, the price of even the smallest authentic adobe would stop your heart.

26. Those weathered old gray stone buildings were painted brilliant red and blue during the Middle Ages.

CITIES

Cities are big places full of buildings full of machines and people full of tension and anxiety and streets full of noisy cars and sidewalks full of crowds and dog droppings. But without the city—the hive of humankind—much of civilization would not exist. Jane Jacobs, an urbanist who has written much on the subject, believes that the city has been more important to human development than the nation-state.

1. What was the first word spoken by a human on the moon?

2. Match these famous ancient cities—now only myth and ruin—with their modern countries or states:

1 Mohenjo-Daro		a	Kampuchea
2 Uruk		b	Sudan
3 Mesa Verde		c	Illinois
4 Thebes		d	Iraq
5 Meroë		e	Syria
6 Zimbabwe		f	Zimbabwe
7 Cahokia Mounds		g	Greece
8 Syracuse		h	Egypt
9 Carthage		i	Pakistan
10 Angkor		j	Colorado
11 Palmyra		k	Sicily
12 Delphi		l	Tunisia

3. What is the oldest city founded by Europeans in the New World?
New York
Mexico City
Havana
Santo Domingo
Saint Augustine

4. What is the westernmost capital city in continental western Europe?

5. How many Moscows are there in the USSR?

6. How many Moscows are there in the United States?

7. What is the second largest French-speaking city in the world?

8. What do Australia, Brazil, and the United States have in common?

9. What city was once called "Florence on the Elbe"?

10 There is only one city inside the Arctic Circle. Can you name it?

11. There is an unusual kind of city called a necropolis. What sort of people would you find there?

12. Define these terms:
a *Glaswegian*
b *Oxonian*
c *Conch*
d *Edokko*

13. Look at the simplified street maps given and name
the city each represents:

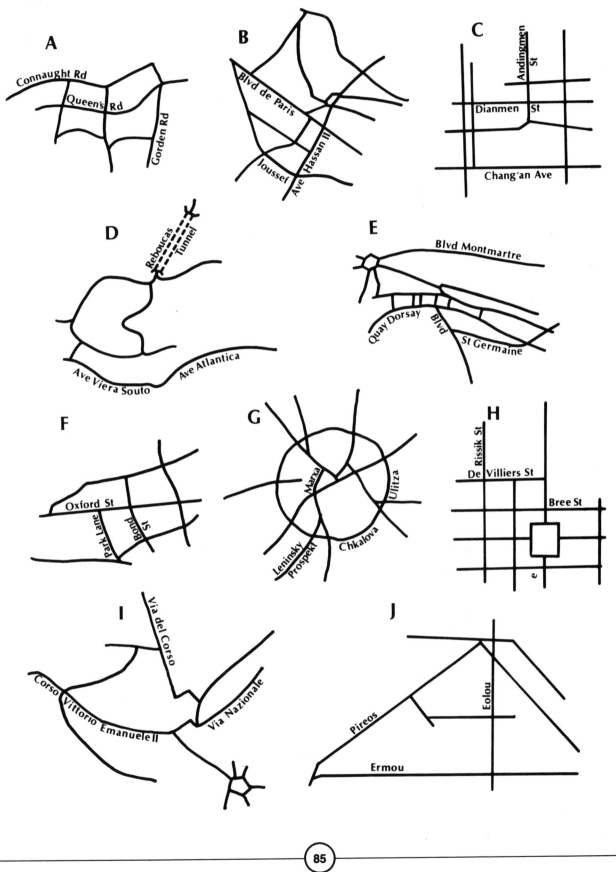

A

Connaught Rd

Queen's Rd

Gorden Rd

B

Blvd de Paris

Joussef

Ave Hassan II

C

Andingmen St

Dianmen St

Chang'an Ave

D

Reboucas Tunnel

Ave Viera Souto

Ave Atlantica

E

Blvd Montmartre

Quay Dorsay

Blvd St Germaine

F

Oxford St

Park Lane

Bond St

G

Marxa

Ulitza

Leninsky Prospekt

Chkalova

H

De Rissik St

De Villiers St

Bree St

I

Via del Corso

Corso Vittorio Emanuele II

Via Nazionale

J

Pireos

Eolou

Ermou

14. Match these parks with their cities:
1 Hyde Park — **a** Mexico City
2 Rock Creek Park — **b** Los Angeles
3 Parque Ibirapuera — **c** London
4 MacArthur Park — **d** Washington, D.C.
5 Stanley Park — **e** Buenos Aires
6 Golden Gate Park — **f** São Paulo
7 Central Park — **g** Paris
8 Lincoln Park — **h** Chicago
9 Bois de Boulogne — **i** New York
10 Bosque de Chapultepec — **j** San Francisco
11 Parke Almirante Guillermo Brown — **k** Vancouver

15. What city has the second largest Chinese community in North America?

16. What capital city was designed on a circular plan centuries ago to symbolize its identity as "the navel of the universe"? Hint: It's not Washington, D.C.

17. Here are the initials of three cities often referred to by their initials. Name the cities:
B.A.
L.A.
P.E.

18. From time to time, entire cities burn. Name the city that burned in:
1666
1906

19. What was a doge?

20. What town or city name would you say is the most frequent in the United States?

21. Cities were once located at the hubs of transport and trade—at the mouths of rivers or on protected harbors. In Europe the Industrial Revolution changed this, and great new cities sprang up near a newly valued resource. What was it?

22. What do the following cities have in common: Canterbury, Moscow, Peking, Rome, Medina, Mecca, Jerusalem, Benares, Gaya?

23. What is a "primate city"?

24. Match these real or legendary capital cities with their real or legendary kings:

1 Aachen — **a** Ramses
2 Camelot — **b** Hannibal
3 London — **c** David
4 Tenochtitlán — **d** Louis XIV
5 Thebes — **e** Julius Caesar
6 Rome — **f** Montezuma
7 Carthage — **g** Queen Elizabeth
8 Jerusalem — **h** King Arthur
9 Paris — **i** Charlemagne

25. A million people died in a battle for this single city during World War II, more than the United States lost in the entire war. It was named one thing then and is called something else now. Give one of those names.

26. These names all have to do with a single famous city: Palatine, Capitoline, Quirinal, Caelian, Aventine, Esquiline, and Viminal. Can you tell what they are, and where?

27. Had you been riding in a satellite three thousand years ago, you might have seen a neat north-south rectangle and perhaps recognized it as the beginnings of a certain city, one of the earliest of those planned on a grid system and probably the only one still surviving—it shows up today in satellite photos. What city is it? Hint: It is not far from the only human-made structure that the astronauts could see from the moon.

28. In northeast India, at 7,500 feet, is a town inhabited chiefly by Nepalese and Bhutanese and famed for its lovely view of the Himalayas—and for its tea. Name that town.

29. How are these cities related to each other: Tokyo, Tehran, Albuquerque, and Memphis, Tennessee?

30. What's common to the distinctive settings of San Francisco, Hong Kong, Nairobi, and Johannesburg?

31. Match these galleries with their cities:
1 Tate — **a** Miami
2 Louvre — **b** Malibu
3 Uffuzi — **c** Paris
4 Guggenheim — **d** Washington
5 Smithsonian — **e** New York
6 Vizcaya — **f** London
7 Getty — **g** Florence

32. What is the windiest large city in the United States?

CITIES

1. "Houston," as in "Houston, *Eagle* has landed," was the first word spoken by man on the moon. This was in 1969, before Houston became one of the fastest-growing U.S. cities, but when it was already the site of NASA's Johnson Space Flight Center.

2.
1–i	7–c
2–d	8–k
3–j	9–l
4–h	10–a
5–b	11–e
6–f	12–g

3. Santo Domingo is the oldest city in the New World.

4. Lisbon, at longitude nine degrees five minutes west, is nearly 200 kilometers west of Dublin.

5. There is one Moscow in the USSR.

6. There are at least six Moscows in the United States. They're in Idaho, Kansas, Maryland, Minnesota, Pennsylvania, and Tennessee.

7. Montreal's population of 3 million is second only to that of Paris (9 million).

8. Australia, Brazil, and the United States have capital cities that were specially planned and located on sites where no city previously stood. This is unusual—most capital cities grow where other cities were.

9. "Florence on the Elbe" was Dresden, lovely and sophisticated until February 1945, when the Allies gave it a saturation-bombing so severe that 150,000 people died and the city was virtually leveled. Kurt Vonnegut was a prisoner of war in Dresden then, but survived to write popular antiwar satires.

10. Murmansk. (There are no cities south of the Antarctic Circle.)

11. Dead people are the sort you'd find in a necropolis, a city of the dead. A necropolis is what archeologists call a huge cemetery in the Middle East; usually the graves are raised above the ground in one way or another.

12. A *Glaswegian* is a person from Glasgow, Scotland; an *Oxonian* is a person from Oxford, England, or one who studied at its famous university; a *conch* is a person born in Key West, Florida (a "freshwater conch" is one who has lived there from earliest childhood and behaves in a properly conchlike fashion—but this title is never more than honorary and provisional); an *edokko* is a person from Tokyo.

13.
   a Hong Kong
   b Casablanca
   c Beijing
   d Rio de Janeiro
   e Paris
   f London
   g Moscow
   h Johannesburg
   i Rome
   j Athens

14.

1–c	7–i
2–d	8–h
3–f	9–g
4–b	10–a
5–k	11–e
6–j	

15. The largest Chinese community in North America is San Francisco's Chinatown, but you knew that, so we didn't ask it. Number two is in Vancouver, British Columbia, Canada, which also features "The World's Thinnest Office Building," only six feet wide.

16. Baghdad, capital of Iraq, founded in A.D. 754 on the ruins of many earlier cities on the west bank of the Tigris, was the "navel of the universe." The "navel" notion was no mere conceit; it was geographical reality. For the next five hundred years, its site permitted Baghdad to control the Arabian Gulf and Mesopotamia, funneling nearly all east-west trade through itself or subjugated ports.

17. The initialed cities are Buenos Aires, Los Angeles, and Port Elizabeth, South Africa.

18. London burned in 1666, San Francisco in 1906.

19. The doge was the ruler of Venice, a person elected from among the ruling families to hold office for life with the nominal approval of an assembly of nobles.

20. The only way to know is to count them, and since atlases vary in comprehensiveness, you can never be sure. Rand McNally's *1985 Road Atlas* has 121 Fairviews, 105 Midways, and 94 Oak Groves. There are also many Clintons, Franklins, Springfields and Greenvilles.

21. New cities grew near Manchester, the Saar, the Ruhr, Lorraine, Silesia, and the Donets Basin, because here were deposits of a newly useful fossil fuel: coal.

22. These cities are all religious capitals. In case you're wondering, Gaya, near Calcutta, is the holiest city of Buddhism.

23. A primate city is not the monkey house. Instead it is, according to geographer Mark Jefferson, a city that is exceptionally large relative to other cities in the country and expresses the national character vividly. Most European capitals are primate cities.

24.

1–i	6–e
2–h	7–b
3–g	8–c
4–f	9–d
5–a	

25. Stalingrad is now Volgograd.

26. These are the seven hills of volcanic rock upon which Rome was built, though not in a day.

27. Beijing, or Peking, in China. More precisely, what you'd see are the walls of the Forbidden City, in the center of the Chinese capital. The nearby feature is the Great Wall, said to be visible from the moon.

28. Darjeeling.

29. At 35°N. latitude they are equidistant from the equator.

30. These are among the very few great cities not located on rivers. Rivers are central to most human settlements. San Francisco and Hong Kong are on rocky headlands with great harbors. Nairobi was founded as a station on the railway from Mombasa to Lake Victoria. Johannesburg is built on a ridge above the world's greatest gold deposit.

31.

1–f	5–d
2–c	6–a
3–g	7–b
4–e	

32. Boston is windiest, with average breezes of 12.5 mph. Chicago is second at 10.3 mph.

TRANSPORTATION

Transportation is geography in motion. It is resources, products, ideas, and people crossing space with the aid of ever-changing technologies. Transportation is Brazilian steel being transshipped to a barge in New Orleans for the journey to become a Volkswagen in Pennsylvania. It's also phoning from your Ohio-built Honda parked on a jammed Los Angeles freeway to order a new battery for your mother's golf cart in Florida. Transportation today is more efficient and cheaper than it has ever been. As a result we see and experience a greater array of goods, services, and ideas than at any other time in history. The scene, in a TV commercial, of an apparently smiling camel riding across the Saudi Arabian desert in a Toyota truck makes this point for us all. Sometimes vehicles, telephones, routes, and freight rates seem to control our lives. Perhaps this is the ultimate price of life in the twentieth century.

1. What was it that first made Lee Iacocca's reputation?

2. Circle the side of the street they drive on in these places:

Japan	L R
New Zealand	L R
Australia	L R
Bahamas	L R
Jamaica	L R
Ireland	L R
United Kingdom	L R
South Africa	L R
Botswana	L R
Indonesia	L R
Korea	L R
Hong Kong	L R
Uganda	L R
U.S.A.	L R

3. The train called *The City of New Orleans* ran between which two cities?

4. What is the last important aviation record? Hint: It has not been set, let alone broken?

5. U.S. Route 66 runs between which two cities?

6. What is the number of California's Pacific Coast Highway?

7. What is freeway phobia and where is it found?

8. a What is the number of Florida's Atlantic Coast Highway?
b Which U.S. highway connects, via bridges, the Florida Keys?

9. Where is the world's largest garage?

10. Where is American Motors' main assembly plant?

11. In what state is the Bonneville Salt Flats, and what do they have to do with transportation?

12. What are "speed bumps" called in Jamaica?

13. What are crosswalks called in England?

14. U.S. Interstate Highways changed North American travel habits. Name the interstate connecting the following pairs of cities:
Miami–New York
Seattle–Portland, Oregon
Chicago–Minneapolis
Detroit–Tampa
Kansas City–Denver
Los Angeles–Phoenix
Dallas–Houston
Saint Louis–Memphis
Boston–Cleveland
Las Vegas–Salt Lake City
San Francisco–New York

15. The United States first mile of concrete highway was laid, in 1909, in a city where it now has a certain double significance. What city was it?

16. What does QANTAS stand for?

17. Where is the world's highest commercial airport?

18. Match the city with its airport:

1	New York	a	Logan
2	London	b	Kimp'o
3	Paris	c	JFK, La Guardia
4	Boston	d	Palam
5	Havana	e	Maiquetia
6	Caracas	f	Kingsford Smith
7	Madrid	g	Dum Dum
8	Calcutta	h	Leonardo da Vinci
9	Delhi	i	Heathrow
10	Seoul	j	Orly, Le Bourget, Charles de Gaulle
11	Sydney	k	Jose Marti
12	Rome	l	Barajas

19. If New York City is on Greenwich Mean Time (GMT) minus 5, and London is on GMT, and the Concorde takes three hours to fly from the former to the latter, and the traveler thereon does not change his or her watch, will the watch read two hours earlier than the real local time when he or she disembarks at Heathrow?

20. To construct a one-hundred-gun, wooden warship in the mid-eighteenth century, about how many acres of trees had to be cut down?
Twenty to thirty
Thirty to forty
Eighty

21. How would you have moved your cattle from San Antonio, Texas, to Abilene, Kansas, in 1866?

22. Sea-Tac and Kai Tak are similar in more than one way. What and where are they?

23. The most luxurious train in the world is found where?

24. Where was Mahatma Gandhi when he first became convinced to fight against British and racial discrimination?

25. The Panama Canal, "the fifty-mile shortcut," now saves mariners a trip of eight thousand miles around the tip of South America. But the Atlantic is about ten feet *higher* than the Pacific, so how does it work?

26. How does the airplane help to save endangered species of animals?

27. How fast does the Ringling Brothers' Human Rocket—the man shot from a cannon—travel?

28. The Wrather Corporation of Beverly Hills tried to purchase a whistle-stop railway station in Wales. What is the name of the station, and why would a show-biz outfit like Wrather want to own it?

29. The original *Orient Express* ran between which two cities?

30. The great 1908 auto race from New York to Paris had a surprising route. What was it?

TRANSPORTATION

1. Iacocca first made his reputation with the success of the Ford Mustang. His idea was that Americans would buy a sporty four-seater if it looked good, had a lot of options, and sold for under $2,400. Millions of us did.

2. They drive on the wrong side of the street in all these places except the United States.

3. Chicago and New Orleans.

4. "The last plum in aviation records," according to Dick Rutan, who hopes to set it, is a flight around the world without refueling. No airplane has ever gotten farther than halfway without a mid-air pit stop. Rutan and Jeana Yeager are building the Voyager, a plane with the wingspan of a Boeing 727 but less weight than a compact car. It will carry nine thousand pounds of fuel—five times its empty weight.

5. Chicago and Los Angeles. Route 66 has been paralleled by a chain of different interstates. The romance, like the 1962 Corvette, is gone.

6. California 1.

7. In southern California, an estimated 10 percent of the average motorist's life is spent on freeways, and some, says research psychologist Roy J. Mathew, have developed an actual phobia. They tremble and sweat on the freeways, occasionally emitting a scream of horror and frustration. Some avoid freeways altogether.

8. **a** Florida A1A
   b U.S. 1

9. Not Los Angeles. It's in O'Hare Airport in Chicago and holds 9,250 cars. (Chicago also has the world's tallest building—the Sears Tower—and the world's largest sewerage works, covering 501 acres.)

10. Kenosha, Wisconsin, where Renaults now roll off the lines that once brought us "Rambler Americans."

11. Utah. Many of the world land speed records have been set there.

12. Sleeping policemen.

13. Zebra crossings.

14. Interstates 95, 5, 94, 75, 70, 10, 45, 55, 90, 15, and 80, in that order.

15. Detroit, where many cars are still built.

16. Queensland and Northern Territory Air Service.

17. La Paz, Bolivia.

18.

1–c	7–l
2–i	8–g
3–j	9–d
4–a	10–b
5–k	11–f
6–e	12–h

19. Yes, the watch will read two hours earlier than local time. What's strange about this is that people seem to take it for granted.

20. To build HMS *Victory* required the felling of eighty acres of oak forest. A smaller, seventy-four-gun ship required the timber equivalent of 3,700 full-grown trees.

22. You'd have driven your herd of cows along the Chisholm Trail to the railhead. The Chisholm was one of four major trails. To the east was the Shawnee, which connected to Baxter Springs, Kansas, and Fort Osage, Missouri. To the west was the Western, which connected Bandera, Texas, with Dodge City, Kansas. And still farther west was the Goodnight-Loving Trail, from Fort Concho, Texas, up the Pecos valley to Denver and into Wyoming. Cattle from southern Texas were also driven to the railhead in New Orleans.

22. Both are airports, one at Seattle-Tacoma, and the other at Hong Kong.

23. The most luxurious train is probably the Blue Train, which runs from Cape Town, to Johannesburg, South Africa three days a week at a sedate 40 miles per hour. It has, among other things, glare-proof windows tinted with pure gold. The service, they say, is wonderful.

24. Gandhi was on a segregated train in South Africa when he decided he'd had enough.

25. Locks are the key to the Panama Canal's operation—twelve of them, each large enough to contain the Eiffel Tower lying on its side. A ship enters, water is pumped in, and the ship rises; a gate is opened, and the ship moves on to the next lock. It is lowered on the other side of the isthmus in the same way. The Atlantic is higher because the earth's rotation floods water into the Caribbean while dragging it away from the Pacific coast.

26. The survival of endangered species, such as the Siberian tiger, may depend on how successfully we can breed them in zoos. But zoo breeding leads to inbreeding (which threatens a species' health and survival), unless care is taken to freshen the gene pool with new animals from elsewhere. The airplane has made this more convenient, as in the Smithsonian's program to ferry three Siberian tigers from the Moscow zoo to the United States, whose Siberian tigers had all descended from a handful of ancestors. Siberians are quite a handful: They are the largest tigers in the world, weighing one hundred pounds more than the Bengal version. They can eat eighty-eight pounds of red meat at a sitting. There are only some three hundred of them left.

27. The Human Rocket hits speeds of seventy-five miles an hour and lands in a net in the same county.

28. Llanfairpwllgwyngyllgogerychwyrndrobwllllantysiliogogogoch is the name. It is thought to be the longest place-name in the world—it certainly is in Britain—and that's why Wrather wanted to own it. Some 200,000 tourists visit the station every year to get their pictures taken under the station sign, buy a ten-inch-long souvenir railway ticket with the station's name on it, or get instructions from the friendly locals on how, with the proper half-dozen pauses, to pronounce it. The sale fell through. The Americans wanted to ship the station to California. Typical, huh?

29. The original *Orient Express* ran from Paris, France, to Istanbul, Turkey. Then it ran back. Of course, there were connections and occasional direct service to London via the ferry.

30. The New York–Paris auto race route was roughly New York City, to Albany, to Chicago, to San Francisco, to Seattle, to Japan, to Vladivostok, to Omsk, to Moscow, to Berlin, to Paris. That's right; they took the long way around. It was an era of few roads, and no gas stations. They drove much of the way on railroad lines, jouncing along the ties. It was the only way to get across Siberia.

HEALTH

The first requirement of health is life. If you are dead, you cannot exercise, eat a proper diet, floss daily, and get regular checkups.

It is easier to stay alive in some parts of the world than it is in others. For example, as this is written, thousands of human beings are dying of starvation in Africa because of a drought. If those same people, with the

same bodies and needs and genetic blueprints, happened to live in Kansas, they would be at little risk.

It is also true that, while there is only one way to be healthy—that is, have little wrong with you—there are lots of different ways to be unhealthy. And some of these, maybe even most of them, have to do with the place you are in.

1. Which American state has the highest rate of new cancers?

2. Whatever became of smallpox?

3. From what you know of the world, match how many persons per doctor you think the following countries have?

1 Ethiopia	**a**	589
2 Mexico	**b**	73,043
3 Japan	**c**	289
4 France	**d**	845
5 U.S.A.	**e**	1,251
6 USSR	**f**	613

4. Food gives health and strength. A healthy young man might consume 3,000 calories in a day. How much TNT would it take to equal the amount of energy he would derive from this?
Two ounces
Six pounds
Twenty-five pounds

5. Where was the worst famine in modern history?

6. The mosquito is the "vector"—or carrier—of five major human diseases. How many can you name?

7. Of the following three countries, which has the highest death rate?
India
Kampuchea
East Germany

8. What clue shows us now, after some seventy years of trying to believe that cancer is not contagious, that it can be?

9. What European country, in the eighteenth century, lost a third of its population to wars and mercenary "services?"

10. Yuko Horikawa is a sociologist who has studied the health of Japanese. He says that because of American-style fast food, Japanese teenagers are changing. How are they changing?

11. For every health worker in the world—nurse, doctor, or dentist—there are two and a half military people. Here are ten countries. Which of them have more military than medical people, and which have more medical than military people?
Japan
U.S.A.
Poland
Switzerland
Ethiopia
Vietnam
Israel
Sweden
United Kingdom
India

12. Kuru is a rare disease, known only in New Guinea. How does the victim get kuru?

13. Place these in order, from highest to lowest life expectancy:
Scandinavia
Japan
U.S.A.

14. Where did syphilis come from?

15. Where did herpes come from?

16. Fisherfolk of Newfoundland and Labrador lacked fresh vegetables during their long winter and survived mainly on white bread. As a consequence they suffered from beriberi and TB, and infant mortality was unacceptably high. In 1944, a law was passed that improved the health of every citizen and gave children a better chance of surviving. What sort of law could do all that?

17. What tropical island has a rate of multiple sclerosis about thirty times higher than you would expect from the size of its population?

18. In what region of the United States are the children fattest? Leanest?

19. A sociologist studied two ethnic American groups to see how their members presented symptoms to the doctor. One of them (group A) tended to be expressive, emotional, expansive—almost dramatic. The other (group B) tended to be stoic and to downplay or even deny their pain. Both start with the letter *I*—can you name the ethnic groups?

20. What is a *traiteur*?

21. What U.S. city or town had the highest rate of Acquired Immune Deficiency Syndrome (AIDS) in mid-1985?
San Francisco, California
Provincetown, Massachusetts
Key West, Florida
Belle Glade, Florida
Cambridge, Massachusetts
Seattle, Washington

22. About how many American women have been raped?
One in ten thousand
One in one thousand
One in one hundred
One in ten

23. Is the U.S. murder rate higher in the North or the South?

24. In what region of the United States do people get the most vitamin C?

25. In what region of the United States is nutrition best?

26. When the Centers for Disease Control in Atlanta surveyed twenty-four states and Washington, D.C., they found that the women of a single state not only smoked the most and drank the most, but also were the fattest. Can you name the state?

27. In the same survey, what state's women were the thinnest?

28. At what time of year do most Americans die?

29. Certain Indians live high in the Andes and over the years have developed certain adaptations to the limited oxygen and low atmospheric pressure. We know of four; how many can you guess or describe?

30. Almost all mammals except humans possess an enzyme called gulonolactone oxidase. What effect does this have on our lives?

31. During this century, we who live in developed countries have grown larger. We have also reached our full growth, and stopped growing, at a younger age—around eighteen or nineteen in the United

States and Europe. At what age did Europeans and Americans reach full size at the turn of the century?

32. When you examine the skulls of ancient Egyptian aristocrats and poor people from the same period, you notice a significant difference in the teeth. What is this difference?

33. The year 1960 was a watershed year in U.S. death-rate statistics. Which statement best describes the situation:

a Before 1960, the death rate was higher in cities; after 1960, the death rate was higher in rural areas.

b Before 1960, the death rate was higher in the country; after 1960, it was higher in the city.

34. London Bridge used to be in London. In fact, in the late 1400s, people lived in 138 houses that had been built on London Bridge. What was the greatest health hazard of passing beneath London Bridge?

35. Nevada and Utah are adjacent states with similar climates. Yet the adult death rate is 40 percent higher in Nevada. Why?

36. What worldwide health problem does the following quotation describe? "Although three-quarters of the population in most developing countries live in rural areas, three-quarters of the spending on medical care is in urban areas, where three-quarters of the doctors live. Three-quarters of the deaths are caused by conditions that can be prevented at low cost, but three-quarters of the medical budget is spent on curative services, many of them provided for the elite at high cost."

37. Delhi belly, Montezuma's revenge, the Pharaoh's curse, Casablanca crud, Hong Kong dog, and Turista are all the same thing. What?

38. Cholera is often transmitted by travelers. Who carried this disease across much of the globe for centuries?

39. What is Pandora's Box?

HEALTH

1. Florida residents have the highest rate of new cancers. This may be connected to the fact that the state has a high population of elderly.

2. Smallpox has been eradicated from the earth—just about. The last known human case was a woman in Africa, who contracted the disease, was treated, and survived. The pox virus itself survives—just in case we need it, presumably—in four laboratories.

3. 1–b   4–f
   2–e   5–a
   3–d   6–c

4. The young man's 3,000 calories is equal in energy to about six pounds of TNT.

5. The worst known famine in modern history took place in China, from 1876 to 1879. Probably 13 million starved. Famines, one will notice, tend to happen where the population is large and the land insufficiently rich to support it. With the spread of deserts and deforestation in the tropical belt—and the growth of human population in precisely those areas of the globe—implications are clear. If you need more than implications, see recent events in Ethiopia and the Sahel.

6. The mosquito carries malaria, yellow fever, dengue (also known as breakbone fever for the terrible pain it causes in the sufferer's joints), filariasis (the presence in the blood of roundworms, carried by mosquitoes), and encephalitis.

7. Kampuchea (formerly Cambodia) had a death rate estimated at nearly 30 persons per 1,000 between 1975 and 1980. Many were murders by the government. The lowest rate has been Samoa, with 3.1 deaths per 1,000 in 1980.

8. Genital warts is what tells us cancer can be contagious. Certain types of wart virus, types 16 and 18, are known to be transmissible from a mother's vulval warts to the baby upon birth, where they take the form of laryngeal papillomas, or warts on the vocal chords, which sometimes become cancerous. In women with some types of genital warts, 28 percent are found to have uterine cervical dysplasia. And 55 percent of women with invasive cervical cancer have one type of genital warts.

9. Switzerland was a nation of mercenary soldiers in the eighteenth century; its modern-day neutrality seems a more successful strategy.

10. Japanese teenagers are ballooning, says Horikawa, on a fattening diet of American fast food. "I'm not just talking about children getting bigger because of the improved diet since World War II," Horikawa says. "I'm talking about obesity—kids who are just plain fat."

11. These have more military people than medical people: U.S.A. (120 per 100), United Kingdom (120), Poland (150), India (260), Israel (670), Vietnam (2,170), Ethiopia (12,330). These have fewer: Japan (50), Switzerland, (30), Sweden (80).

12. Kuru, a malady known only in New Guinea, is a slow, degenerative disease of the brain, much like the disease scrapie in sheep. Science does not know the cause, but it is transmitted by cannibalism—specifically, by eating raw human brains.

13. Japan has the highest life expectancy in the world; the Japanese woman has the longest life expectancy of all humans. Scandinavia is second, the United States third.

14. Columbus's sailors got syphilis from the Indians. They later shared it in Europe.

15. As the wife of a famous rock singer was once heard to observe in the Chart Room bar in Key West, Florida, "I wish to hell I knew!" No one knows. Herpes simplex types I and II have been around for as long as we can tell. It grew more common, though, with the pill and with changing sexual habits. Many people stopped using condoms and spermicides, and it is now believed these were helping to kill or impede the virus. "It used to say 'for prevention of disease only' on condoms," says Dr. Harvey Blank, a dermatologist who specializes in these viruses. "They didn't know it was true."

16. The law required adding vitamin A to margarine and enriching white bread with calcium, riboflavin, thiamine, iron, and niacin. In under ten years, beriberi disappeared. Tuberculosis declined by 81 percent, and infant mortality by 63 percent.

17. Key West. It still remains unexplained.

18. The children of a nine-state area of the Northeast are fatties, according to a study by Drs. William Dietz and Steven Gortmaker, of Tufts University, and those of a seventeen-state area west of Kansas the leanest. The researchers sampled 7,119 children by measuring the fatness of a pinch of flab. Twenty-two percent of the northeastern kids were obese, 18 percent of the midwestern kids, 15 percent of the southern, and about 14 percent of the western. They don't know why. They speculate that ranch activity keeps kids lean, but note that city kids in all regions are more like others in their region than kids in other regions. All over the country, though, they found kids fatter in winter than in summer.

19. Irish Americans were the stoics—the researcher saw their stoicism as a defense mechanism against their "oppressive sense of guilt." The dramatists were Italian Americans. They were seen as coping with anxiety by "repeatedly over-expressing and thereby dissipating it." All this is not merely interesting; it can affect how well a doctor can diagnose disease. A doctor from a stoic culture might dismiss an Italian's complaints as hypochondria, while a doctor from an expressive culture might overlook entirely the symptoms of a stoic.

20. A *traiteur* is a Cajun folk doctor. It means something closer to "treater" than to "traitor."

21. Belle Glade, Florida, a tiny, impoverished farming community of only sixteen thousand people, most of them migrant workers, at the lower edge of Lake Okeechobee, has the highest per-capita rate of AIDS in the world, with thirty-four cases in mid-1985. Half the victims have died. Public-health physicians descended upon the unfortunate town seeking a reason. Researchers remain unclear as to the cause. Transmission via prostitutes from the Caribbean may be an explanation. The squalid living conditions of the farm workers may also contribute as well.

22. According to a study by the Urban Institute in Washington, D.C., about one in ten American women has been raped—a two- to four-times-higher percentage than in western Europe. An additional 10 to 20 percent have experienced rape attempts, the study found. More than half took place indoors, and 70 percent occurred without the use of weapons.

23. There are more murders per capita in the American South than in the North. But when the figures are adjusted for education and income, there is no difference; the South has more poor, ill-educated people, among whom murder is more frequent. Most American murder victims are, as the University of Maryland sociologist Colin Loftin puts it, "dirt-poor people who have been killed by dirt-poor people." But this murder business is complicated. Take the element of race, for instance. If you are a white woman, your chance of being murdered in the U.S. is 1 in 369, according to Justice Department figures. If you're a white man, it is 1 in 131. But if you are a black woman, it is 1 in 104, and if you are a black man, you have 1 chance in 21 of becoming a murder victim. American blacks have a very high murder rate, as both victims and perpetrators. But that is not a racial characteristic: African blacks have a much lower rate than American whites. Nor does it correlate strongly with mere poverty—black Haitians are poorer than U.S. blacks, and their murder rate is low too. Clearly, there are factors at work that we cannot understand or are unwilling to deal with.

24. People in the Northeast get the most vitamin C.

25. Westerners are the best-nourished Americans, according to a study by Lana Hall and Karen Bunch of Cornell University. Westerners eat better than residents of the rest of the country, except for Vitamin C.

26. Florida, where 30.5 percent of the women were overweight, outchubbing the women of Washington, D.C., by a slender margin.

27. California women. Only 13.7 percent were overweight.

28. The peak period of American death is December 15 through February 15, when it is cold, icy, slippery, and depressing.

29. The Indians of the Andes have compensated by developing:

   a Larger-than-average lungs, so they can inhale more air

b Continuous dilation of the tiny lung sacs, so that the surface area of the oxygen-absorbing tissues is increased

c A heart that is up to 20 percent larger than average and is much more muscular, for ease of pumping

d About two quarts more oxygen-carrying blood than coastal people

30. Without this enzyme, we cannot synthesize vitamin C, which means that if we are deprived of it in our diet, we get scurvy. Other animals don't.

31. Americans and Europeans reached full size at about age twenty or twenty-one at the turn of the century.

32. The teeth of aristocratic Egyptians show tartar, plaque, abscesses, and cavities at about the same rate as ours today. Those of the poor people, who ate a diet of coarse and often uncooked food, are much better.

33. The death rate had always been higher in the cities until about 1950, when the situation began to reverse itself, a reversal that was complete in a decade. Ever since 1960, the ratio of rural to urban deaths has steadily increased. You're safer in the city.

34. One of the great advantages of occupying a house on London Bridge in the 1400s was ease of waste disposal. In the rest of the city, open drains were the method. But passengers on open boats plying the Thames had to look sharp when passing beneath, for human waste carries many diseases.

35. The death rate is different from Nevada to Utah largely because people live differently in these states. The contrast could hardly be sharper: the sedate, nonsmoking, nondrinking Mormons in Utah versus the hard-drinking, hard-smoking, up-all-night, risk-taking high-rollers in Las Vegas and Reno. Of course, not everyone in Nevada is a hell-raiser, and not everyone in Utah a stick-in-the-mud, but the stereotype holds true enough to make a statistical difference. One study a few years ago found that, "for the most clearly alcohol- and tobacco-related diseases, liver cirrhosis and lung cancer, the differential is as high as 600 percent."

36. This quote relates to the problem of equal access to health care. Almost nowhere in the world does the physician serve the people equally.

37. These are all names for traveler's diarrhea, or gastroenteritis.

38. Moslems making a pilgrimage to Mecca transmitted cholera across large parts of the globe for centuries.

39. Pandora's Box is where all the diseases of the earth came from. Prometheus stole heavenly fire from Zeus, who became inflamed with rage. To punish the human race, he enclosed all the earthly ills in a box. Diabolically, he placed Hope inside as well and gave the box to a woman named Pandora. She opened it. The diseases escaped, but Hope remained inside. How do we stuff the diseases back into the box and release Hope?

ITEMS OF VALUE

Productwise, as they say, life was not always so rich and interesting. Most products were once available only where they grew or could be picked up, perhaps with a little scratching or digging. If you lived there, you got the product, but not the product from where other people lived. Later, people traveled and met people from other places who had other products, and so they traded back and forth and a new way of making a living arose. It was called trading—swapping, buying, selling. From that, of course, grew a lot of other ways of earning a living. All

this depends, however, not only on the products themselves, but upon how we humans decide to feel about them. A resource is only one because humans perceive it to be one, use it as one, and say so. Sometimes this is because we have found use for an item, such as foodstuffs or building materials. Other times we simply stipulate, perhaps because of beauty or rarity, that such and such a thing, useful or not (gold, for instance), is a valuable resource.

1. Look at the earth as the rich and wondrous thing it is, and say what you think are its two greatest, most marvelous, most valuable resources.

2. Where was the first-ever shopping mall built?

3. Match these cars with their country of original manufacture:

1	Toyota	**a**	Yugoslavia
2	Mercedes-Benz	**b**	U.S.A.
3	Peugeot	**c**	Japan
4	Jaguar	**d**	Sweden
5	Avanti	**e**	Czechoslovakia
6	Alfa-Romeo	**f**	Australia
7	Holden	**g**	Italy
8	Skoda	**h**	United Kingdom
9	Volvo	**i**	France
10	Yugo	**j**	West Germany
11	Bentley		
12	Ferrari		

4. What commodity, after fossil fuel, has the highest total value in world trade?

5. When the Standard Oil monopoly was broken up early in this century, what five states got their very own Standard Oil companies?

6. If you wanted to make some extra money by clubbing baby harp seals to death, where would you have to go to do it?

7. What is GUM?

8. Can you describe the difference between a carrot, a caret, and a karat?

9. Name some countries where people spend the following:
Dollars
Pesos

Pounds
Francs
Shillings
Guilders
Kroner
Rubles
Zlotys
Drachmas
Rupees
Rands
Pulas
Qwachas
Escudos
Bolivars
Yuan
Yen

10. What country exports the most beef?

11. What country imports the most beef?

12. What country produces the most of each of the following?
Barley
Cacao
Rice
Wine
Cane sugar
Citrus
Bananas
Peanuts
Pigs
Cattle
Copper
Iron ore
Chrome ore
Maize
Wheat
Rye
Tea
Oats
Coffee
Cotton
Sheep

13. What country exports the most wool?

14. What country cuts the most wood each year?

15. What country exports the most wood and wood products?

16. What country has the greatest petroleum reserve?

17. Residents of what country eat the most chocolate?

18. What country *makes* the most chocolate?

19. Can you describe what cacao looks like on the tree?

20. What country makes the most silk?

21. How did the South American rain forests influence European furniture design in the late eighteenth century?

22. What kind of grass may have inspired the design of the Chinese junk?

23. Every year, humans use about 2.5 billion cubic meters of wood. A little over half of this is used for building and other industrial applications. What is the other 46 percent used for?

24. Some countries have coal, and others don't. If those that don't have it are at all developed, they need it and therefore must buy it.
Who buys most of Australia's export coal?
Who buys most of the United States' export coal?

25. Which country imports the most oil?

26. Immigrants came to the United States for many reasons, but one was economic opportunity. About how much did they have to invest for a ticket from Bremen to New York, in steerage, in 1837?
$2
$16
$225

27. Where would you guess is the world's biggest McDonald's?

28. Where would you guess is the world's biggest shopping mall?

29. On the island of Yap, pronounced "wop," the natives once used a form of money whose coins were four feet across and weighed several hundred pounds. Of what were these coins made?

30. What is a billion?

31. In which county would you find:
a Disneyland
b Walt Disney World

32. What is the main difference between a Japanese handsaw and a Western handsaw?

33. If you set aside lunacy and thrill factors, the idea of robbing banks is to carry off enough cash to avoid further work for a meaningful length of time. Lots of factors combine to make this a risky business—cops and guns being only two of them. In Bolivia, for example, an unfortunate economic factor called _____ makes robbing banks more trouble than it is worth. Fill in the blank and explain.

34. Which of these peoples would you say saves the most money, relative to their income:
Germans
Swiss
Americans
Japanese

35. It was in Europe—in England, to be precise—that for the first time in history, anywhere, more people began to live in cities than in the country. Why was this point reached, and in about what year did it happen?

36. Who has the Bomb?

37. The United States has the biggest advertising industry in the world, as you might expect. Which country has the second biggest?

38. Where was Cheddar cheese invented?

39. Where was the paisley-print design developed?

40. You might assume that the United States and the USSR are the world's largest arms exporters, and you'd be right. Who is the third largest?

41. Where is the kitty-litter capital of the world?

42. Where is the last place on earth still making steam locomotives?

43. Are the "developing countries" developing?

44. Three countries are the top investors in U.S. industry. Japan just joined that group by becoming third. Can you name the first two?

45. Match the country with its per-capita gross national product. This is a figure representing the total value of goods and services a country produces, divided by its number of people. The average of the whole world is $2,754 (1983 figures).

1	U.S.A.	a	$10,330
2	USSR	b	4,170
3	Japan	c	25,850
4	Venezuela	d	8,950
5	Mexico	e	12,530
6	India	f	253
7	Kuwait	g	4,701
8	United Kingdom	h	2,250

46. What are the following? And where are they?
Harrod's
Gump's
Rich's
Bloomingdale's
Neiman-Marcus
Marshall Field
Burdine's

ITEMS OF VALUE

1. Probably the richest and most important resources on earth are the oceans and the tropical rain forests. Both have supported and nurtured humans, and today both are in danger of being systematically ransacked and poisoned.

2. The first-ever shopping center was built in the mid-1950s, in Valley Stream, Long Island, New York.

3.
1–c		7–f
2–j		8–e
3–i		9–d
4–h		10–a
5–b		11–h
6–g		12–g

4. Coffee has the second highest value in world trade.

5. Indiana (Standard of Indiana, known as Amoco now), Ohio (Standard of Ohio, known until recently as Sohio, now Standard Oil Company), California (Standard of California, now called Chevron), New Jersey (Standard of New Jersey, known as Esso, a name that was changed to Exxon), and New York (Standard of New York, now known as Mobil).

6. To club baby seals, take your Louisville Slugger to one of the four harp-seal whelping grounds: the Magdalen Islands in the Gulf of Saint Lawrence, the southeast coast of Labrador, the ice floes north of Jan Mayen in the Greenland Sea, and the ice floes of the White Sea. But think first.

7. GUM is the giant state-run department store in Moscow.

8. A *carrot* is a long, slender, orange-colored vegetable with greens on one end and a point on the other. It is high in carotene, which the liver converts to vitamin

A. Eating too many carrots can make one's skin turn an unearthly orange. Jewelers use the *karat* as a unit of weight measurement—a one-karat gem weighs 200 milligrams. It is also the measure of purity in gold alloy—something that is one karat is one-twenty-fourth pure. Thus, twelve-karat gold is 50 percent gold; twenty-four-karat gold is pure gold. A *caret* is a printer's symbol to indicate the place where something is to be inserted.

9. People spend dollars in the United States, Canada, Hong Kong, the "British" Caribbean, Australia, New Zealand, and a lot of other places; pesos in Mexico and several other Latin American countries; francs in Belgium, France, and Switzerland, as well as several countries in west and central Africa; schillings in Austria and most of East Africa; guilders in Holland; kroner in Denmark; rubles in the USSR; zlotys in Poland; drachmas in Greece; rupees in India and Pakistan; rands in South Africa; pulas in Botswana; qwachas in Zambia; escudos in Portugal; bolivars in Venezuela; yuan in China; and yen in Japan.

10. Australia exports the most beef. Argentina is next.

11. The United States imports the most beef. Italy is next.

12. Barley, USSR, China next; cacao, Ivory Coast, Brazil next; rice, China, India next; wine, Italy, France, next; cane sugar, India, Brazil next; citrus, United States, Brazil next; bananas, Ecuador, Costa Rica next; peanuts, India, China next; pigs, China, USSR next; cattle, India, United States next; copper, United States, USSR next; iron ore, USSR, Australia next; chrome ore, South Africa, USSR next; maize,

United States, China next; wheat, USSR, United States next; rye, USSR, Poland next; tea, India, China next; coffee, Brazil, Colombia next; cotton, United States, USSR next; sheep, Australia, USSR next.

13. New Zealand.

14. USSR, United States next.

15. Canada exports the most wood.

16. Saudi Arabia has the world's greatest petroleum reserve, and Kuwait is next.

17. The Swiss lead the world in consumption of chocolate, devouring twenty-two pounds per person every year. Americans eat only two to ten pounds per year, despite a large number of true chocolate addicts.

18. The United States actually makes the most chocolate, and imports the most. The greatest portion comes from Ivory Coast, and from a single state in Brazil: Bahia. Together, these produce 45 percent of the world's cacao.

19. The cacao tree produces ribbed, melonlike fruits along its trunk. Inside these are white, beanlike seeds in a thready pulp. The beans, when dried, treated, and ground, become the raw material for chocolate.

20. China still produces more than half the world's silk.

21. When Europeans realized that mahogany's dense, close, even grain permitted the carving of thin, tapered legs, furniture became much more delicate of line.

22. Bamboo, a grass, may have inspired the junk's design. The structure of the junk, a series of independent waterproof compartments, mimics that of the bamboo stalk. This notion did not appear in Western naval architecture until very recently—in, for instance, the *Titanic*.

23. Forty-six percent of the wood consumed by humans every year is simply burned up as fuel within a few miles of where it was cut.

24. Japan buys most of Australia's export coal, and western Europe buys most of the United States'.

25. The United States imports more oil than any other country, with Japan a close second. Japan has no oil deposits of its own.

26. Steerage passage to New York cost $16 in 1837. This was several weeks' wages then.

27. The world's biggest McDonald's is on the Pacific Island of Guam, which is only twenty miles long. "It's as big as a K mart," says the historian Rick Panedero, who has been there.

28. The biggest shopping mall in the world apparently is in São Paulo, Brazil, and it is large enough to contain a supermarket with 140 checkout counters staffed twenty-four hours a day. There's another huge one in Edmonton, Alberta.

29. The Yap coins were made of stone.

30. In the United States, a billion is one thousand million. This is perverse, and worth bearing in mind when you exchange cash at airports, because everywhere else in the world, a billion is one million million. This is why it is so much easier to become a billionaire in America.

31. a Disneyland is in Orange County, California.
b Walt Disney World is in Orange County, Florida.

32. The Japanese handsaw cuts on the pull-stroke, the Western on the push-stroke.

33. Hyperinflation. In Bolivia, the inflation rate was 8,216 percent or so in mid-1985, making the bank robber's greatest problem one of carrying away a worthwhile amount of cash. As we write, a one-hundred-pound bag of Bolivian one-thousand-peso notes is worth three hundred dollars. It makes good mulch, they say.

34. The Swiss save the most—at least there's more per capita in Swiss banks, which may not be the same thing. Relative to their income, the Japanese are far ahead, followed distantly by the Germans, who are followed by the Americans.

35. Beginning around 1851, more English were living in cities than in the country, because the Industrial Revolution had changed the way most made their living.

36. It's difficult to tell who has the bomb, beyond the obvious: United States, USSR, England, France, China, and India. Israel may have it. Pakistan, bitter enemy of India, may have, since Pakistani agents have been caught buying technology from around the world that adds up to a bomb. Iran may also have it, or be on the verge of having it. The bombs known to be in existence are equivalent to five thousand times the power of all the munitions of all kinds used by all sides in the entirety of World War II.

37. Japan has the second biggest advertising industry in the world, gaining a full percent of its GNP from it. The biggest spenders are the food and beverage industries.

38. Cheddar cheese was developed in Cheddar, England, where the key variable was the aging of the cheese in the cool, even temperature of nearby limestone caves.

39. Paisley designs were developed in Paisley, a city in southwest Scotland, famous as an eighteenth-century textile center.

40. France is the third largest exporter of arms, having sold $7.1 billion worth in 1984, a record. Most went to Saudi Arabia and other Gulf states.

41. The kitty-litter capital of the world is Quincy, Florida, where there are kaolin mines. Kaolin is an absorbent type of clay.

42. The last country making steam locomotives is the People's Republic of China, where a factory in Datong, near the Manchurian border, was recently turning out one a day for the P.R.C.'s vast network of railways. The factory is being phased out.

43. The developing countries may not be developing. At least that is the implication of one disturbing statistic. The average annual percentage of export sales growth of developing countries was 7.8 percent in the 1960s. During the 1970s, however, it had fallen to 1.4 percent. This may suggest that, relatively, the have-nots are falling further behind, a possibility that should be worrisome to wise haves. As a group, the African countries have been doing the worst.

44. Japan is the third heaviest foreign investor in U.S. business. Canada is the second, and the United Kingdom the first, with about $3 billion a year. Others are the Netherlands and West Germany. Japan's primary motive, it is said, is to insulate itself from U.S. protectionism by operating *within* the United States.

45.

1–e	5–h
2–g	6–f
3–a	7–c
4–b	8–d

46. These are department stores headquartered in London, San Francisco, Atlanta, New York City, Dallas, Chicago, and Miami, respectively.

GEOGRAPHY AND LANGUAGE

People invented language so that they could communicate with each other. But that was before humans had cars or airplanes, or for that matter had even domesticated the horse, so different languages developed in different regions.

And similar languages diverged when the people using them moved apart. French, Spanish, and Italian all stem from a common ancestor, Latin.

Language became a source of identity and pride. The French, who have established a national academy just to try to keep foreign words like *drugstore* and *weekend* out of French, are perhaps the most absurd example of linguistic hubris. Humans being what they are—that is, human and not French—words do jump from one language to another, sometimes with curious results. Just look at English.

1. Translate the following from Spanish, Russian, and Japanese, respectively, into English:
   a *Gua-gua*
   b *Af-af*
   c *Wan-wan*

2. Where is Geographe Bay?

3. In what country is each of the following languages spoken?

1 Amharic	**a** South Africa
2 Basque	**b** East Africa
3 Mandarin	**c** Ethiopia
4 Farsi	**d** West Africa
5 Kanarese	**e** Spain, France
6 Zulu	**f** Iran
7 Yoruba	**g** China
8 Tungus	**h** USSR (Siberia)
9 Georgian	**i** USSR (Caucasus)
10 Swahili	**j** India

4. How should you pronounce Pago Pago?

5. Define these common words from the English language:
   a *Tripper*
   b *Buffer*
   c *Spanner*
   d *Coster*
   e *Sultana*

6. Eskimos have a lot of words for snow. About how many *do* they have? And why?

7. How many words do *we* have for snow?

8. Bedouins in Saudi Arabia have about 160 words for what?

9. What American word has covered the most world territory?

10. Where does the term *Yankee* come from?

11. What and where are the boondocks?

12. What language does *honcho* come from and what does it mean?

13. What do Regina, Saskatchewan, Canada, and Key West, Florida, U.S.A., have in common?

14. Name some U.S. state capitals that have men's first names.

15. Four U.S. state capitals were named for presidents; can you name them?

16. At least two U.S. state capitals were named for foreign honchos; can you name them?

17. Name some world capitals that have women's names.

18. Name some world capitals that have men's names.

19. Among whom are clicking sounds an integral part of language?

20. When Europeans arrived in the New World, about how many languages were spoken there?
 a Thirty to fifty
 b Three hundred
 c Two thousand

21. Match the nationality with the usual content of its dirty jokes:
 French — Oral sex, the abasement of women
 American — Homosexuality, incest
 English — Seduction, adultery, sex technique
 German — Pee-pee, doo-doo

22. What are *ticos*?

23. The citizens of a newly formed country are often of many tribes or nations, speaking many languages. Nigeria, for instance, contains speakers of two hundred different native tongues. How do these people communicate with one another?

24. When Europeans invaded the New World, they not only killed or enslaved most of the people, they tried to extirpate the existing culture—some of which was superior to their own. They burned, for instance, virtually the entire written history of the Aztecs. In many ways they succeeded, but a few ancient languages are still spoken today by surviving Indians. Which of these is among them?
 Navaho
 Nahuatl
 Maya

25. Most "pidgin" languages are based on European languages, because as colonizers Europeans have been present at the times of great political, military, or economic ferment that produces pidgin tongues. In Chinese-English pidgin, what does the word *pidgin* mean?

26. Translate this from the Texan: "Ahminna awl bidniz."

27. About what proportion of the world's population is bilingual?
 4 percent
 25 percent
 50 percent
 80 percent

28. Name a few countries that have a single, uniform national language.

29. In most countries, you might live in a *casa,* or *maison;* in an English-speaking country you'd live in a _____. But much of the time you'd refer to it not as "my _____" but as "my _____," a word that connotes a great deal about your feelings for the place in which you live. In most languages, however, including Spanish and French, there is no equivalent word. This might be illuminating. Or it might not. Fill in the blanks.

30. In what country was the first book set in moveable type?

31. Translate the following from French and German into English:
 a *Le Jaseroque—Il brilgue: les toves lubricilleux Se gyrent en vrillant dans la guave*
 b *Der Jammerwoch—Es brillig war. Die schlichte Toven Wirrten und wimmelten in Waben*

32. Match the now-English word with the language from which we borrowed it:
 1 *Alcohol*   **a** Japanese
 2 *Yen* ("craving") **b** Hebrew
 3 *Boss*   **c** Hindi

4 *Kayak*   **d** Irish
5 *Emotion*   **e** Yiddish
6 *Amen*   **f** French
7 *Punch*   **g** Arabic
8 *Smithereens* **h** Chinese
9 *Rocket*   **i** Dutch
10 *Tycoon*   **j** Eskimo
11 *Kibitzer*   **k** Italian

33. What is the most widely spoken language on earth?

34. What was the lingua franca?

35. Which country has the highest percentage of adult illiterates?
Nepal
U.S.A.
Costa Rica
Nicaragua
Poland
Bolivia
Philippines

36. In what country would you hear the term *sheila,* and to what would it refer?

37. Draw a horizontal rectangle or a square. Then draw a vertical line down through the center of it. Translate this ideograph into English and tell what country it refers to.

38. To what does "coon-ass" refer?

39. Fill in the blanks with the adjectival form of a nation:

_____ jumping beans
_____ fly
_____ horn
_____ windlass
_____ bacon
_____ cheese
_____ checkers
_____ crawl
_____ two-step
_____ measles
_____ delight
_____ muffin
_____ fries
_____ Dry
_____ standoff
_____ flu
_____ bank account
_____ nut
_____ lawyer
_____ towel

_____ pastry
_____ treat
_____ Shave
_____ twins
_____ carpet
_____ rug

40. It is strange but true that by straining a little, a Japanese, a Chinese, and a Korean can stand side-by-side and read a sign written in any of the other's languages, but could not read the same sign to one another over the phone and be understood. Please explain.

41. How did the Bahamas get their name?

42. What English word did these two Chinese words give us: *ku li*?

43. What is toponymy the study of?

44. How do you pronounce the following:
Dominica
Anguilla
Grenada (the Caribbean island)
Grenada (the city in Spain)

45. Dar el Beida is the largest city in Morocco, but it is often known by another name, which has come to embody the exoticism of the Mideast. What is that other name? Hint: Parts of Morocco were occupied by the Spanish.

46. There is more than one United States in the Western Hemisphere. There are three. Can you name them?

47. Where is the Cape of the Eleven Thousand Virgins?

48. You're probably like us. When you meet someone, you want to know their name. When you see something new, you want to know the word for it. Wouldn't it be strange to be someplace and not have a name to call it? Of course, you'd not be there long before you started calling it something. But what would you call it? Well, there are at least five ways that places get their names. Can you name them?

GEOGRAPHY AND LANGUAGE

1. **a** Bow-wow
    b Bow-wow
    c Bow-wow
2. Geographe Bay is an inlet on the southwestern coast of Western Australia.
3. Amharic is the language of the Ethiopian highlands. Basque is spoken on both sides of the Pyrenees, in Spain and France. Mandarin is the official language of China. Farsi is the official language of Iran. Kanarese is one of the many languages of India. Zulus are the largest ethnic group in South Africa. Yoruba is one of the major languages of Nigeria in West Africa. Tungus is a language of Siberia. Georgian is spoken in Soviet Georgia. Swahili is spoken in East Africa.
4. Say "pango pango."
5. These British English words mean, in American English:
    a "tourist"
    b "bumper"
    c "wrench"
    d "pitchman" (barker, salesman)
    e "raisin"
6. Eskimos have about twenty different words for snow. Languages reveal what is important to people, how carefully they have observed them, and how intimately they live with them. In English we have only one word for water, whether it's in an ocean, a bathtub, or a Thermos jug. Hopi Indians have a different word for water in a lake and water in a jug. English has only two words—*this* and *that*—to show something that is here or there. Some Indians, such as the Tlingits of the Northwest Coast, have one word for something near and always present, another for something a little farther away but still always present and near, another for a thing even farther away, and still another for objects out of sight. That sort of information must matter to a Tlingit. The Masai, who live on cattle, have seventeen words for them. And look how many words we have for "money."
7. We have, according to *Roget's Thesaurus*, only about six, and they don't really discriminate very usefully between types of snow, the way an Eskimo's words would. They are *snow, snowdrift, snowflake, snowstorm, snowball, snowslip*. There must be more—we can think of *snowfall* and *snowman*—but snowwise, the Eskimos have it.
8. The Bedouin have 160 words for camel, which incorporate descriptive details of age, sex, color, and ancestry.
9. Probably *okay*, or *OK*. It's been used virtually everyplace American servicemen have been since the days of Andrew Jackson, and for better or worse that's just about everyplace. It was Jackson, linguists say, who popularized the word first, having taken it from the Choctaw word *okeh*.
10. Manhattan. The word *Yankee* was originally *Jan Kees*, a derogatory term for the cheese-eating Dutch (Jan Kees means "John Cheese"). The Dutch turned it to *Yankee* and stuck it on their neighbors of English descent.
11. "Out in the sticks" is usually what it means, of course, though its meaning was more specific when the word was *bundok*, meaning "mountain" in Tagalog, the language of the Philippines. The word

probably came to the United States with soldiers from the 1898 Spanish-American War. Warfare spreads language and culture among its survivors.

12. *Honcho* may sound Spanish, but it is Japanese. It means "army squad leader," approximately, but in the United States it has come to mean any sort of boss or leader.

13. Regina and Key West have little in common except their names. Regina's started out, in 1882, as Pile of Bones. Key West's original name was in Spanish: Cayo Hueso, which means "Bone Key" or "Isle of Bones" ("Key West" would be Cayo Oeste).

14. Pierre, Austin, Montgomery.

15. Madison, Lincoln, Jackson, Jefferson (City).

16. Bismarck, Raleigh.

17. Sofia and Victoria are two; can you think of more?

18. There seem to be few capitals with men's names. Santiago is the only one we can think of, and we're not certain of it. Or of the meaning of all this, although men do most of the naming.

19. The Bushmen of the Kalahari use up to four different kinds of vocalized clicks as consonants. Several other languages use one click as a consonant.

20. When Europeans arrived, they found people who spoke about 2,200 different languages, more than were known in all of Europe and Asia combined. These were not "primitive" tongues, either—there is no such thing as a primitive language. All languages serve all of their speakers' purposes. There are some seven thousand words used in the King James Bible. The typical Sioux or Aztec employed ten thousand words in everyday conversation.

21. In our experience, German jokes tend toward pee-pee, doo-doo themes; the French laugh at seduction and adultery; the English giggle over homosexuality and incest; and the Americans titter about oral-sex themes and the abasement of women.

22. Ticos are persons from Costa Rica, so called for their habit of adding *tico* instead of *tito* to words to make diminutives, as in *momentico*, meaning "little moment."

If a *momentito* is a "little moment," does that mean a *burrito* is a little burro and a *mosquito* is a little mosque?

23. People in polyglot countries like Nigeria usually must communicate officially in the language of their former colonial oppressor—English, in this case. Forty-one of the forty-nine African nations have chosen English, French, or Arabic as an official language, ignoring their various native tongues. Even the Somali Republic, which does have a native language spoken by more than 90 percent of its citizens, has declared Arabic, Italian, and English as official languages—but

not Somali. Ethiopia is an exception. That country has a long, proud history of independence and a language of its own: Amharic. Only a bare majority of Ethiopians, however, are native speakers.

24. All three of these languages—Navaho, Nahuatl, and Maya—are still spoken.

25. *Pidgin* is the word for "business."

26. "I'm in the oil business."

27. "Well over half of the world's population achieves the bilingual switch from one language system to another scores of times each day," writes Peter Farb in *Word Play*.

28. No country has a single, uniform national language. Even France, so paranoid about its mother tongue that it maintains a national institute of linguistic purity, cannot succeed and has speakers of Basque in the Pyrenees, German in Alsace-Lorraine, and Provençal in the south.

29. *house, house, home*.

30. Johannes Gutenberg had nothing to do with it. It was in China that a book called *Dreamland Essays* was set in type of wood and clay—later editions used metal type—in the eleventh century A.D. The type printed ideographs, of course, not letters.

31. Both are versions of lines from Lewis Carroll's nonsense poem, "Jabberwocky":
> *'Twas brillig, and the slithy toves*
> *Did gyre and gimble in the wabe . . .*

32.

1–g	7–c
2–h	8–d
3–i	9–k
4–j	10–a
5–f	11–e
6–b	

33. About 726 million speak Mandarin Chinese, most of them in China. But 397 million or more speak English every day, and another 400 million employ it as a second language. Britain's centuries as a colonial and military power helped to spread the language, and the influence of the United States continues the process. The richness and flexibility of the language itself helps. About 274 million speak Russian, 254 million speak Hindi, and 251 million speak Spanish.

34. The original lingua franca was a hybrid language of French, Italian, Spanish, Greek, and Arabic and was spoken in Mediterranean trading ports during the twelfth to the fourteenth centuries. The phrase now, by metaphor, means "any hybrid language used for communication between different peoples."

35. In Nepal, 99.6 percent of the twenty-five-year-olds have no school education, making it the least literate country on earth.

36. In Australia, a *sheila* is a young woman, though the word is not always a polite reference to a young woman. It means something like the Americanism "broad."

37. The ideograph you drew is Chinese. A square represents the world, the line in the center represents the location of China in the world and thus stands for what Chinese call their huge and ancient land: the Middle Kingdom. The phrase connotes a China that is the center of the universe around which all other "uncivilized" activities revolve.

38. "Coon-ass" is what you'd probably better not call a Cajun, of the Louisiana bayous, whose farmer ancestors came from southern France to Nova Scotia in the early seventeenth century and founded Acadia. When the British arrived one hundred years later and drove them out, many Acadians settled in Louisiana, where the locals, having already twisted their name from Acadian to Cajun, now risk calling them coon-asses. They don't like it.

39. Mexican jumping beans, Spanish fly, French horn, Spanish windlass, Canadian bacon, Swiss or American cheese, Chinese checkers, Australian crawl, Aztec two-step, German measles, Turkish delight, English muffin, French fries, Canada Dry, Mexican standoff, Asian flu, Swiss bank account, Brazil nut, Turkish towel, Danish pastry, Dutch treat, Burma Shave, Siamese twins, Persian carpet, Oriental rug.

40. Chinese, Japanese, and Koreans all use the same language of pictures—some fifty thousand of them, somewhat modified over the years by divergence—but when they speak, give them different sounds.

41. The Bahamas is a chain of some seven hundred islands spread across 100,000 square miles of shallow sea and coral reef in the Atlantic Ocean (not the Caribbean Sea, as some tourist brochures claim). *Baja mar* in Spanish means "low sea" or "shallow sea."

42. The Chinese words *ku li*, meaning "bitter strength," gave us the English word *coolie*.

43. Toponymy is the study of place-names.

44. Dom-in-EEK-a, An-GWILL-a, Gren-AY-da, Gren-AH-da.

45. The other name of Dar el Beida is Casablanca, "white house" in Spanish.

46. The United States of America, Los Estados Unidos Mexicanos, *y* Estados Unidos de Venezuela.

47. The Cape of the Eleven Thousand Virgins is the headland at the eastern end of the Strait of Magellan. There are many islands here, each standing for a virgin.

48. Here are five ways we know of that places get named:
   a They're named after people, as in Johnstown.
   b They're named for how they look, as in Green Valley.
   c They're named in memory of some other place, as in New Buffalo, Michigan.
   d They're named after an event or curious discovery that took place there, as in Broken Arrow.
   e They're named in transliteration from some other language, as in Miami.

III

WORLD REGIONS

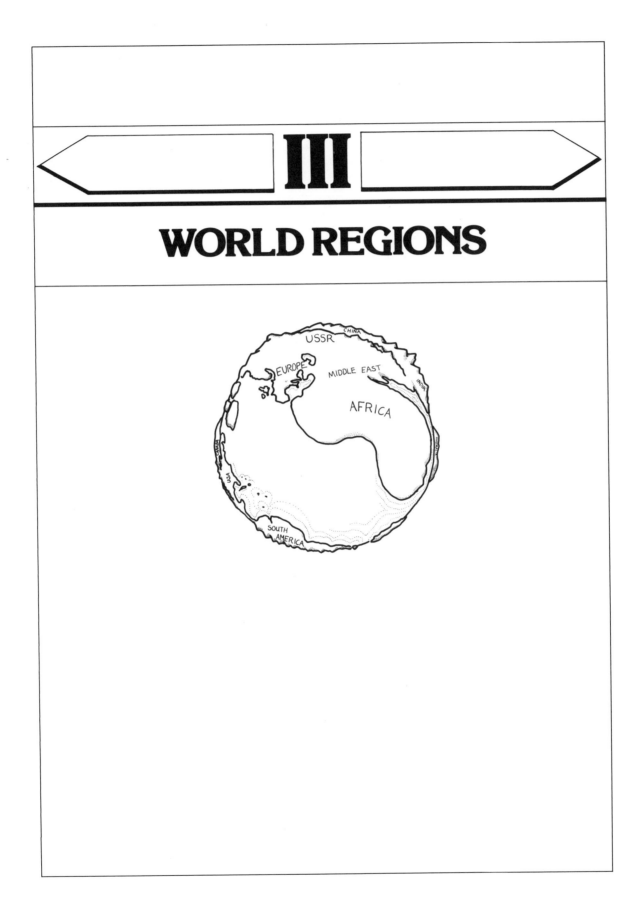

EUROPE

Europe, once, was where most North Americans' ancestors came from. Now, many of us come from other places: Asia and Latin America and the Middle East, to name three. But our culture is still largely a European one, as are our ways of thinking. As you travel, you can see how the arts, for instance, began in the Middle East long ago, how they solidified in Egypt, were copied and refined and enhanced and, you might say, Westernized by the Greeks, were then enriched, somewhat softened, and politicized by the Romans, and spread westward, finally, across the Atlantic. The languages we speak came from Europe. And so, even if our family has lived in North America for generations, and even if they came here from some other place than Europe, we're all in a way "from" Europe—part of our character derives from the European experience. Maybe that's why Europe is where most North Americans go when they vacation abroad—to see where they came from.

"Can we never extract this tapeworm of Europe from the brain of our countrymen?" Emerson asked in 1860.
Probably not.

1. On the lid of the Häagen Dazs ice-cream container there is a map of Scandinavia, with two cities labeled: Oslo and Copenhagen. Where is this ice cream with a cult following actually manufactured?

2. Match the descriptive with its country (thanks to Luigi Barzini):

The Imperturbable	Dutch
The Quarrelsome	Americans
The Flexible	British
The Mutable	French
The Careful	Germans
The Baffling	Italians

3. If New Caledonia is in the South Pacific, where is "old" Caledonia?

4. In this quotation from Paul Theroux, identify the country: "There were no blank spaces on the map of _____, the best-known, most fastidiously mapped, and most widely trampled piece of geography on earth."

5. A British hundredweight weighs more than a hundred pounds. Can you say why?

6. Here are some ancient, probably faintly familiar place names: Bosnia, Herzegovina, Montenegro. They're all in the same country now. Which one?

7. Give an answer to this old droll geographer's riddle: When you're racing across northern Europe, how do you know where the finish line is?

8. In what western European country are three languages officially spoken, and what are they?

9. Bullfights are held in two European countries.
   a Name those countries
   b How do their bullfights differ?

10. History buffs know where Alexander the Great came from.
  a Where?
  b Where is that place now?

11. Where in the world does Norway find itself situated between Mexico and China?

12. Napoleon won at Austerlitz and lost at Waterloo—in what countries are these places now?

13. Fjords along the coast of Norway were formed by ancient glaciers. Are they deep or shallow?

14. There are three main language groups in Europe, into which most of the national languages fall. Name the language groups of:
French, Spanish, Portuguese
German, Dutch, Swedish
Polish, Czech, Russian

15. Compared with other regions on the same latitude, western Europe has a very moderate climate. Why?

16. Name two landlocked countries in western Europe.

17. You know that Europe is bounded on the west by the Atlantic, north by the Arctic Sea, and south by the Mediterranean. But what is its eastern boundary?

18. Two of present-day Europe's most important countries were unified from many small states only in the last century. Can you name them?

19. What and where is Flanders?

20. In 1800, in Britain, about 9 percent of the people lived in cities or towns. A hundred years later *(25 percent, 62 percent, 80 percent)* lived in towns. Pick a percentage and explain why.

21. When we speak of the British Isles, we're speaking chiefly of two islands: Britain and Ireland.
Which is larger?
Which lies closer to the mainland?

22. If you instructed Einstein on matters of relativity, or shipped air conditioners to Nome or bananas to Guatemala, it would be tantamount to carrying _____ to _____.

23. How did the United Kingdom get hold of Ireland, Northern Ireland, Scotland, and Wales?

24. Why are Belgium, the Netherlands, and Luxembourg called the Low Countries?

25. The most densely populated independent nation in the world is a certain European principality in which the people speak French and whose defense is handled by the French. What is it?

26. What sort of places are the Ruhr and the Saar?

27. What is a canton?

28. Which country is a leading producer of false teeth?
Liechtenstein
France
San Marino
Austria

29. What island is just off the toe of the Italian boot, and what social-financial organization has it given the world?

30. One of the smallest nations in the world is also the oldest country in Europe, and the oldest republic in the world. Most of its citizens earn their living making and selling postage stamps. Name it.

31. These people live in the Pyrenees Mountains between Spain and France and have been there longer than either the Spanish or the French. Their language, which they call Euskara, appears to be unrelated to any other language in the world. They are the _____.

32. The Douro River flows through rich vineyard lands and emerges north of Lisbon at the port city of _____, from which much _____ wine is exported.

33. It has endured many invasions and wars, became part of the eastern Roman Empire between 197 B.C. and the mid-fifteenth century, was then conquered and ruled by the Turks for four hundred years before becoming a free country again in 1829. For many of the years since, its people have lived under dictatorships, either by occupying powers or by their own military. This country is _____, the birthplace of _____.

34. The rain in Spain falls mainly:
   a On the plain
   b Somewhere else

35. Frigg, Alwyn, Piper, and Ekofisk are curious new European mini-cities. What do they have in common?

36. Match the following well-known regions with their countries:

1	Algarve	**a**	Austria
2	Andalusia	**b**	Yugoslavia
3	Calabria	**c**	Romania
4	Finnmark	**d**	Denmark
5	Jutland	**e**	Italy
6	Transylvania	**f**	France
7	Serbia	**g**	Spain
8	Macedonia	**h**	Portugal
9	Thrace	**i**	Netherlands
10	Tyrol	**j**	Greece, Yugoslavia
11	Bavaria	**k**	Norway
12	Alsace	**l**	Greece, Bulgaria
13	Friesland	**m**	Greece
14	Peloponnese	**n**	West Germany

37. Where is the "pearl of the Adriatic"?

38. A country some people joke about today was once the largest in western Europe. Which country, and when was this?

39. Five large European countries remained neutral throughout World War II. Name them.

40. About how far west did Attila get with his Huns before he turned around and went back east?
   Italy
   France
   England
   Ireland

41. The Roman Empire was sensibly divided into provinces. Place these provinces in their modern countries:

1	Cappadocia	**a**	Libya
2	Dacia	**b**	Greece
3	Dalmatia	**c**	France
4	Gaul	**d**	Turkey
5	Numidia	**e**	Romania
6	Achaea	**f**	Yugoslavia

42. On what river is Cambridge, England?

43. The U.S. president has a country retreat at Camp David, in the Maryland Hills. Where is the British prime minister's country house?

44. Match the church with its city:

1	Saint Peter's	**a**	Paris
2	Saint Mark's	**b**	Rome
3	Saint Paul's	**c**	London
4	Notre Dame	**d**	Venice

45. Match the "real" names of these European countries with their English names:

1	Sverige	**a**	Albania
2	Norge	**b**	Greece
3	Magyar	**c**	Poland
4	España	**d**	Finland
5	Suisse	**e**	Ireland
6	Deutschland	**f**	Switzerland
7	Ellás	**g**	Spain
8	Eire	**h**	Hungary
9	Suomi	**i**	Norway
10	Polska	**j**	Sweden
11	Shqiperia	**k**	Germany

46. How many armed robberies would you say have occurred in the history of Iceland?

47. Probably all children play with dolls, have dollhouses, and generally rehearse themselves for the domain of the grown-up. Not all of them get to play with something like the world's greatest dollhouse, which is built on a scale of one inch to a foot and includes working electrical appliances and lights—and is found where?

48. In the United States, when we shake hands, we pump one or two times—anything over three raises questions. How many pumps of the hand are characteristic of the European handshake?

EUROPE

1. Häagen Dazs is made in Teaneck, New Jersey.
2. In *The Europeans*, Barzini's chapter titles are: "The Imperturbable British," "The Quarrelsome French," "The Flexible Italians," "The Mutable Germans," "The Careful Dutch," and "The Baffling Americans." (He considered Americans to be an odd, latter-day form of European.)
3. "Old" Caledonia is in the North Atlantic—it is another name for Scotland.
4. Great Britain, in *The Kingdom by the Sea*.
5. In the United States, a hundredweight equals 100 pounds, although the term is seldom used. In Britain, weight is often measured in stones rather than pounds (perversely, the English measure their *money* in pounds), and since a stone equals 14 of our pounds, and a hundredweight is composed of eight stones, it weighs 112 pounds.
6. Yugoslavia.
7. You have crossed the Finnish line when there are no more Lapps.
8. Flemish, a language like Dutch, is spoken primarily in northern Belgium, French is spoken primarily in southern Belgium, and German is spoken along the eastern border of Belgium. Switzerland has four official languages: German, French, Italian, and Romansh.
9. a Spain and Portugal.
   b In Spain, they kill the bull.
10. a Alexander the Great came from Macedonia.
    b It is now in Greece and Yugoslavia.
11. Norway, the eleventh country to be "represented" at Walt Disney World's Epcot Center, in Orlando, Florida, has been inserted neatly between the equally fake representations of China and Mexico.

12. The Austerlitz at which Napoleon won is in Czechoslovakia; there's another in upstate New York. The Waterloo at which Napoleon lost is in Belgium; there's another in Iowa.
13. Fjords are very deep, some nearly four thousand feet deep. They were formed by the erosion of glaciers far below present-day sea level during the ice ages.
14. The first are Romance languages, the second are Germanic, the third are Slavic. Within each group, all the languages derived from a common ancestor, but different ways of speaking evolved over centuries of isolation, very much as birds isolated in the Galápagos Islands evolved differently from their cousins elsewhere.
15. The Atlantic Ocean prevents most of western Europe from experiencing extremes of temperature. Westerly winds blow inland off the ocean, keeping winters mild and summers cool.
16. No part of western Europe is farther than three hundred miles from the ocean, and there aren't many landlocked countries there. Four of the larger landlocked ones are Austria, Switzerland, Hungary, and Czechoslovakia. The mini-states of Luxembourg, San Marino, Andorra, Liechtenstein, and Vatican City are also inland.
17. The precise location of Europe's eastern boundary has always been in dispute, among geographers as well as warring nations who lived on it. Some say Russia is European and that the Urals form the boundary between Europe and Asia. Others place the line farther west, claiming that because eastern Europe and western Europe are both fragmented into small countries, each different from the other, and Russia is an immense monolith, the boundary should be there,

between Russia and eastern Europe.

18. Italy began to coalesce after the Napoleonic Wars (1815) and became the Kingdom of Italy in 1870. Germany was unified by Prussia in 1866.

19. Flanders, important as a trading center in the Middle Ages, became famous again as the scene of bloody trench warfare during World War I. It was never a country in the modern sense, but merely a region in what is now parts of Belgium and northern France.

20. Because of the Industrial Revolution, 62 percent of the English lived in towns by 1900, and 90 percent do today.

21. Britain is both larger and nearer to the mainland, only twenty-one miles away.

22. Coals to Newcastle, the port for the coal mines of northeast of England, which of course had plenty once. Today, U.S. coal is carried to Newcastle, which has run low.

23. Ireland is not part of the United Kingdom; it stands alone as Eire, or the Republic of Ireland. The others, for the most part, wish they weren't, all of them still having independence movements of various strengths. Wales was conquered by the English in the Middle Ages. A Scottish king ascended to the English throne in the seventeenth century, bringing Scotland with him. As for Northern Ireland, it was kept by the British when it surrendered the rest of Ireland in 1921 because a large number of Scottish and English Protestants had settled there. Most of Ireland, of course, is Catholic.

24. Because they are low. Almost half of Holland lies below sea level.

25. Monaco, on the French Riviera.

26. Both the Ruhr and the Saar are heavily industrial areas. The Ruhr is Europe's largest industrial region, containing the cities of Düsseldorf, Dortmund, and Essen. The Saar, near the French border west of the Rhine, produces wine as well as steel.

27. A canton is a governmental unit in Switzerland, which has twenty-three of them. Each canton has more power to govern itself than the national government has to govern it.

28. Although Liechtenstein has fifty-some industries and is known for its low taxes on business corporations, its second greatest income-producer is the sale of postage stamps to foreign collectors. It is also a leading European producer of false teeth.

29. Sicily, the rugged, mountainous island birthplace and exporter of La Cosa Nostra—the Mafia—is just off Italy's boot-tip.

30. San Marino, on the slopes of the Apennines entirely within Italy, has been an independent republic since 1631. It covers twenty-four square miles.

31. They are the Basques.

32. Oporto exports port.

33. Greece, the birthplace of democracy.

34. The rain in Spain hardly falls on the plain at all. The high plains, or *meseta central*, of central Spain are actually semiarid. Most of the rain falls along the far northwest coast, beyond Portugal.

35. Frigg and the rest are all in the North Sea. They are oil and gas fields halfway between Norway and Scotland.

36.

1–h	5–d	9–l	13–i
2–g	6–c	10–a	14–m
3–e	7–b	11–n	
4–k	8–j	12–f	

37. The pearl of the Adriatic is in Yugoslavia: Dubrovnik.

38. From the Peace of Westphalia in 1648 until 1667, Poland extended from near the Oder River in the west to include nearly all of what is now the Ukraine and White Russia in the east. The second largest country in western Europe then was the Kingdom of Sweden.

39. Switzerland, Sweden, Ireland, Spain, and Portugal all remained neutral during World War II.

40. Attila and the Huns went as far west as France, to a place then called Aurelianum, probably in A.D. 450. Aurelianum is now Orléans.

41.

1–d	4–c
2–e	5–a
3–f	6–b

42. Cambridge is on the Cam.

43. Chequers, the prime minister's getaway spot, is a Tudor manse in Buckinghamshire, thirty-five miles northwest of London.

44.

1–b	3–c
2–d	4–a

45.

1–j	4–g	7–b	10–c
2–i	5–f	8–e	11–a
3–h	6–k	9–d	

46. In the entire history of Iceland, there has been one armed robbery. Iceland has only 200,000 people, but even so . . .

47. The Queen's Dollhouse is in Windsor Castle. It was built in the 1920s for Queen Mary (1867–1953). Besides the lights and appliances, there is a miniature wine cellar filled with great little vintages, musical instruments, and a library full of leatherbound books, many contributed in the handwriting of the original authors.

48. Europeans pump the hand about five to seven times. Anything less may be regarded as chilly or aloof.

EUROPE

EUROPE

Locate the following on the accompanying map of Europe (countries are marked with letters, cities and islands with numbers, rivers and mountains with double letters):

London _____

Paris _____

Spain _____

Norway _____

West Germany _____

France _____

Italy _____

Iceland _____

Greece _____

Danube River _____

Hungary _____

Rome _____

Crete _____

Sweden _____

Oslo _____

Zurich _____

Romania _____

Po River _____

Carpathian Mountains _____

Milan _____

Amsterdam _____

Dublin _____

Helsinki _____

Sicily _____

Poland _____

Marseilles _____

Belgium _____

Lisbon _____

Bonn _____

Loire River _____

Munich _____

Bordeaux _____

Tiranë _____

Andorra _____

Sofia _____

Belgrade _____

Vistula River _____

EUROPE MAP QUIZ

London **3**

Paris **31**

Spain **I**

Norway **B**

West Germany **M**

France **H**

Italy **W**

Iceland **A**

Greece **Z**

Danube River **bb**

Hungary **R**

Rome **26**

Crete **23**

Sweden **C**

Oslo **5**

Zurich **34**

Romania **S**

Po River **cc**

Carpathian Mountains **ff**

Milan **35**

Amsterdam **10**

Dublin **2**

Helsinki **7**

Sicily **27**

Poland **P**

Marseilles **32**

Belgium **K**

Lisbon **30**

Bonn **12**

Loire River **dd**

Munich **16**

Bordeaux **33**

Tiranë **21**

Andorra **37**

Sofia **22**

Belgrade **36**

Vistula River **gg**

ASIA

Asia is, simply stated, impossible to comprehend. Its size, mass, numbers, antiquity and complexity defy the human grasp. This region has the world's two oldest continuous civilizations (India and China); it includes the world's largest federal state (India); it has some of the most dramatic physical geographic variety and most intense cultural fragmentation; it represents achievements so impressive and problems so profound that we can only, in frustrated despair, ask you:

1. Where did Donald O'Connor find Francis the Talking Mule in the first "Francis" movie (1950)?

2. Why don't the efficient Japanese use word-processing computers in business, the way managers in most industrial countries do, instead of sending handwritten letters and memos to each other?

3. During what years did the United States last occupy Japan?

4. About how many sumo wrestlers are there in Japan, and how much do they weigh?

5. What city was the capital of the British East India Company?

6. When you go over the Khyber Pass, from what to what do you pass?

7. You would find these—Sindhi, Baluchis, Punjabi, and Pathans—in Pakistan. What are they?

8. Who was the Guru Nanak? What is the *Granth Sahib*?

9. How did Pakistan, which sounds to us so strange and foreign and Asian, get its name?

10. There are five landlocked countries in Asia—name two.

11. Japan is made up of many islands, but most of its territory is in the four main islands. Name two.

12. Where was the third Disney theme park built?

13. Occidentals have been the world's champion colonizers and imperialists, but some orientals were imperialists too. Which Asian country was the most so?

14. You might guess that the world's most populous conurbation is in the Orient; if so, you're right. Which is it?

15. The rebellion called the Meiji Restoration did something special for Japan in 1868—what was it?

16. Most of the refugees fleeing Indochina in the late 1970s—400,000 of them—came to the United States. But almost three-fourths as many went somewhere else. Where?

17. A man who later became president of the United States found himself defending his wife and other Stanford University alumni during the Boxer Rebellion in Tientsin, China. Who was he?

18. Sri Lanka, like many another country, is ethnically divided in two, with a minority that believes itself oppressed. The two groups are the Sinhalese and the Tamils. Which is the minority?

19. What was the Dutch East Indies and what was its capital?

20. What city has been called "the Venice of Asia"? Hint: It played a part in the James Bond movie *Man with the Golden Gun.*

21. In what country is our year 1986 called 62?

22. In which city is the Taj Mahal?

23. What is the last course in an authentic Chinese meal?

24. About 35 percent of Americans attend a university. What proportion of Chinese go on to college?
About 3 percent
About 11 percent
About 25 percent
About 35 percent
About 50 percent

25. In one Asian language, Chinese, there is no word for "privacy," and no word for "intimacy." But there is a word—*renao*—that does not have a counterpart in English or any European language. Its literal meaning is "hot and noisy." Can you give a rough translation?

26. The Chinese love to read; why is there a shortage of books in China?

27. A type of porcelain is named after an Asian country. What is it?

28. Who said, "The enemy advances, we retreat; the enemy camps, we harass; the enemy tires, we attack; the enemy retreats, we pursue"?

29. In the sixteenth century, the Portuguese brought the gun to a certain Asian country, which since then has had a reputation as both warlike and good at stealing other country's ideas. What country was it, and what did it do with the gun, which it learned to manufacture as well as the Portuguese?

30. In Japan which member of the family is usually in charge of money?

31. Where are the following?
Causeway Bay
Repulse Bay
Aberdeen
Happy Valley
Victoria Peak

32. If you spent a week in Beijing, about how many dogs would you see on the streets:
One or two
Perhaps one hundred to one thousand
Many more than one thousand

33. In Chinese, how do you pronounce:
Qin
Ge
Xian
Zedong
Zhou

34. Match the animal with the sound the Japanese people say it makes:

1	Cat	a	*Hee-hee-hin*
2	Dog	b	*Gao-gao*
3	Mouse	c	*Moh-moh*
4	Pig	d	*Wo-wu,* or *wan-wan*
5	Cow	e	*Niao*
6	Duck	f	*Choo-choo*
7	Horse	g	*Boo-boo*

35. About how many McDonald's are there in Japan?
Four
Forty
Four hundred
Four thousand

36. Until World War II, sake was the most popular alcoholic beverage in Japan, but beer slowly caught up and surpassed it. During the last ten years, something else has surpassed beer—what is it?

37. Where is the Ginza, and what does this word mean?

38. Why is it easy for a stranger to get lost in Tokyo?

39. What does the Great Wall of China protect? About how long is it?

40. How do you tell a Chinese person from a Japanese person?

41. Who are the Han people?

42. Name the only art form that survived China's Cultural Revolution almost intact.

43. Where did the Boxer Rebellion get its name?

44. What Asian country has had such good crops that it exported 500,000 tons of grain to the USSR and even gave 100,000 tons of grain for African famine relief in 1985?

45. Fill in the blanks of this caption from a *New Yorker* cartoon, in which a little girl is reading a report to her classmates: "This past weekend, my parents took me to see *The King and I. The King and I* is a musical version of *Anna and the King of Siam.* Siam is a country that is now called _____. _____ is next to Kampuchea. Kampuchea is a country that was once called _____."

46. How many of Japan's Big Six automakers can you name?

47. South Asia encompasses India, Pakistan, Sri Lanka, and Bangladesh. It reaches from the slopes of the _____ in the north to the island of _____ in the south.

48. Match the dominant religion with the country:
1 Sri Lanka	**a**	Hinduism
2 India	**b**	Buddhism
3 Japan	**c**	Islam
4 Pakistan	**d**	Shintoism

49. Between 1857 and 1947, when India became independent, the British government ruled that country as a colony. What body ruled India in the period that ended in 1857?

50. In 1983, Bangladesh had 1,750 persons per square mile of total land, and India had 430. These figures are the arithmetic population densities of these two countries, and from them it is clear that India is somewhat better off than is Bangladesh. But if you compare a more significant figure called the physiological population density, things look even worse for Bangladesh. What does this figure represent?

51. India is often depicted as a place of teeming cities. About what proportion of Indians actually live in urban areas?
15 percent
25 percent
45 percent
65 percent

52. When the English colonists checked out of India in 1947, they left their railroads behind. This was useful to the Indians, of course, and that country's railway system is the world's fourth largest. But even today, it causes enormous amounts of extra work. Why?

53. India's population at the turn of the century was 250 million people, and experts believe it had remained that size for centuries before, fluctuating with disease, famine, climate, and war. What caused India's dramatic population increase?

54. Bangladesh was part of another country until 1972. What country was it part of?

55. Where is Chittagong?

56. Pakistan keeps moving its capital, its locus of government function. First it was Karachi. Then the country's political functions were moved to Rawalpindi, and later to Islamabad. Why?

57. In what Asian country are the majority of the people Aryans?

58. In all the world, there is only one Hindu monarchy. Which country is it?

ASIA

1. Burma. Francis must have been unique. Not a single talking animal has been found there since.
2. Not oriental politeness. The Japanese language uses ideographs, not letters, and there are about two thousand of them in common use, most with more than one meaning. Figuring out how to put them all on the usual typewriter keyboard is a serious problem. The need to make Japanesse computer-compatible may change totally the way the language and writing are taught.
3. The United States occupied Japan most recently during 1945–52.
4. Japan has seven hundred registered sumo wrestlers. Some are very weighty indeed, but according to regulations, they need weigh only 165 pounds.
5. Before the British came to India, Calcutta was three little fishing villages on a coastal mud flat, founded in 1696, very late by Indian standards. It was the capital of British India until 1911.
6. Crossing the Khyber Pass, you travel from the Indus Valley in Pakistan to the high plateau country of Afghanistan, following the channel of the Kabul River.
7. These are the four major ethnic groups of Pakistan.
8. Guru Nanak, born in the mid-fifteenth century, founded the Sikh religion, a blending of old Hindu and Islamic religions, now itself one of the world's great religions. Sikhs live mostly in the Punjab of northwest India and eastern Pakistan. The *Granth Sahib* is their holy book.

9. Pakistan was named by the British, who also invented it before they checked out of this part of Asia in 1947. They named it, cleverly, by taking a *p* from *Punjab*, an *a* from *Afghanistan*, and a *K* from *Kashmir* and put them all together with *stan*, which means "land or place of."
10. Asia's landlocked countries are Afghanistan, Mongolia, Nepal, Bhutan, and Laos.
11. Japan's four main islands, from north to south, are Hokkaidō, Honshū, Shikoku, and Kyūshū.
12. Mimicry being a fundamental Japanese cultural trait, the third Disney theme park was built, in 1983, just outside Tokyo.
13. Japan has been the Orient's major colonizer. With few resources at home except a smart, proud, and hardworking people, the island nation has often used its need for resources as justification for aggression—against other Pacific islands, against China, and even against the United States, in World War II. Since the war Japan has been so successful at buying raw materials from distant sources, adapting advanced technology to the production of fine goods, and selling those goods worldwide at great profit, that some observers have wondered whether the whole thing was planned as a Hundred Year War of world economic conquest.
14. Tokyo-Yokohama-Kawasaki has more than 23 million people; one of every four Japanese lives there. Greater Mexico City is about 4 million behind, but trying hard to catch up.

15. After three hundred years of turning inward, Japan's Meiji monarchy was reestablished. Control was vested in a group of revolutionary leaders determined to bring Japan to the status of a world power, and Japan's modern period began.

16. China.

17. Herbert Hoover, a mining engineer by training, was prospecting in northern China in 1900 when the Boxer Rebellion began, and found himself leading a defense of this European outpost near Beijing.

18. The Tamils are the minority in Sri Lanka.

19. Batavia was the capital of the Dutch East Indies. Since the independence of Indonesia in 1949, it has been known as Jakarta.

20. Bangkok—remember the chase scene along and over the canals?—is "the Venice of Asia."

21. With the accession of each new emperor, the Japanese begin their calendar over with the Year One. In 1986, Hirohito had reigned for sixty-two years.

22. The Taj Mahal is in Agra, the ancient Mogul capital of India. The world's most beautiful building is a mausoleum, built in the middle 1600s. Today, a Holiday Inn is within walking distance.

23. Soup is the last course in a Chinese meal.

24. About 3 percent of Chinese go to college.

25. "Hot and noisy" suggests, as Fox Butterfield describes it, "the pleasure and excitement of a large group of friends and relatives who get together for a meal, with everyone talking and plenty of lights in the room." The Chinese cannot understand what we would call a quiet, intimate dinner, with the lights down low—they prefer to eat with noisy people in bright, hot, noisy rooms.

26. China has a shortage of books because it has a shortage of trees from which to make paper. Peasants denuded the country for firewood centuries ago with little replanting, and the shortage was exacerbated in recent years by the voluminous publication of propaganda put out during the Cultural Revolution. Even so, many clerks and elevator operators today can be seen reading books when not otherwise employed, a rare sight in most Western countries.

27. China.

28. Mao Zedong said it, used it, and "liberated" his country from the Nationalists with it.

29. The country was Japan, as you knew, and what it did with the gun was simply reject it. A period of peace had arrived in the island nation, and the Japanese saw no reason to build weapons for which they had no clear and immediate need.

30. Although Japan is largely a male-dominated society, the wife is usually in charge of the family funds, even to the point of disbursing weekly pocket money to the husband.

31. All these are in Hong Kong.

32. You'd see only a handful of dogs in Beijing. The Chinese government had them all killed in the early 1980s, and continues to discourage their presence in the cities. Also, many Chinese favor a good boiled dog on a cold winter's evening. At the Ye Wei Xiang (Wild Fragrance) restaurant in Beijing, a house specialty is Fragrant Dogmeat Stew, traditionally prepared only from the flesh of young black dogs.

33. In the new Romanization of the Chinese language, *qin* is pronounced "chin," *ge* is "grr," *xian* is "shee-an," *Zedong* (Mao's given name; last names come first in Chinese) is "tse-doong," and *zhou* is "joe."

34. 1–e
2–d
3–f
4–g
5–c
6–b
7–a

35. There were, at last count, 402 McDonald's restaurants in Japan.

36. Whiskey is the most popular alcoholic drink in Japan these days, followed by beer, followed distantly by sake. Together these two now account for 90 percent of the alcohol drunk in Japan. (Sake is of course served warm, but must not be heated hotter than 122 degrees F, or the flavor will be ruined.)

37. The Ginza is a section of Tokyo, now a kind of Times Square. *Gin* means "silver," and *za* means "place." This was the neighborhood in which silver was minted between 1600 and 1800.

38. It's easy to get lost in Tokyo because this is one of the biggest, most populous cities in the world and has grown with virtually no zoning, so that factories stand next to houses next to schools next to bars next to ancient Shinto temples, and there are almost no street names, numbers, or signs. The best way to tell a stranger how to find you is to memorize the locations of coffee shops in various parts of the city, rehearse giving directions to them, and then describe their location to your visitors. Then you can go and meet them there.

39. The Great Wall is not so great anymore, if its protective function is all that's considered. But for a time, it did serve to protect China from the hordes of Mongolians who lived to the north. A 1984 survey discovered 2,500 more miles of it, for a total of 6,200 miles in several branches. The Great Wall, seen close up, truly engages the imagination.

40. There is no certain way to tell a Chinese from a

Japanese. After all, the Japanese probably came from China originally. And China, although most of its people are of the Han group, has fifty-six other ethnic groups, each of which has a different language and, some might say, a different look. But you can make an educated guess, based on the principles offered to one of the authors by an observant Chinese named Mr. Qi: Japanese tend to resemble each other more than do the Chinese, and they are shorter, paler, and have larger faces. It sounds weird, perhaps, but it works more than half the time, a pretty good average when you are dealing with human beings, who refuse to be stereotyped this way.

41. The Han, one of China's fifty-seven ethnic groups, amount to 94 percent of the Chinese people. The others—the Manchurians, Mongolians, Tibetans, and Uygurs, to mention a few—occupy just under two-thirds of China's land, much of it important border territory.

42. Calligraphy survived the Chinese Cultural Revolution, and samples of Mao Zedong's own calligraphy are shown proudly throughout the country. Calligraphy, done well, combines art with scholarship.

43. The Boxer Rebellion got its name from the Society of Righteous Fists—Yi He Tuan—set up to fight European encroachment in China. Europeans called its members the "Boxers."

44. India actually exported grain in 1984. But this does not mean that that overpopulated land is licking its own food problems, except in isolated enclaves such as the wheat fields of the Punjab, where the exported grains came from. Yield per acre is extremely low in India, and as the population grows, the amount of cultivated land per person declines. Thus the problem will get worse, unless something is done quickly. Little land reform was carried out when the British left, and even into the late 1970s, 4 percent of the country's farming families owned a full third of its cultivated land.

45. Thailand. Thailand. Cambodia.

46. Japan's Big Six automakers are Toyota, Nissan, Isuzu, Mitsubishi, Honda, and Mazda.

47. Himalayas. Sri Lanka.

48. 1–b
2–a
3–d
4–c

49. India was ruled by a private corporation, the East India Company, until the populace rebelled in 1857 and the British government took over.

50. The physiological population density represents the number of persons per unit area of *productive* land.

In these terms, India does not look healthy—it has 325 persons per square kilometer. But Indians might surmount this problem. Bangladesh's situation seems hopeless; its physiological population density is four times greater than India's, and there is less than a fifth of an acre of productive land per person.

51. Only about a fourth of India's people live in cities. But that amounts, in a place as populous as India, to 180 million humans.

52. The English railroads in India were built by various companies, each according to its own ideas of railroading. Today, four different gauges—the distance between tracks—are in use, so that to ship goods across the country may require loading and unloading four different trains.

53. The arrival of Europeans caused India's population to spurt. The Europeans brought improved food production and distribution, better sanitation and medicine, and more control over larger territories so that brushfire wars were reduced. The result was a lower death rate. But the birthrate remained the same, with the result we see today: 140 million more Indians in the last ten years. A plan of compulsory sterilization of persons with more than three children was instituted in the mid-1970s, and nearly 4 million were sterilized, but citizens rioted and the program failed. There are more than 20 million sterilized citizens in India.

54. Bangladesh was part of Pakistan until 1972, when a war of independence set it free. It was called East Pakistan then, and its partner (now called the Islamic Republic of Pakistan) was West Pakistan. They were on opposite ends of India and had in common only their Islamic faith, which was not enough. This Florida-sized nation has 96 million people and fertile soil and, before the war of secession, was able to produce 80 percent of its own food. But during the war, fields and crops were abandoned to rot, and now the Bangladeshis are falling ever further behind. One in five depends on foreign-aid food imports.

55. Chittagong is Bangladesh's port city, on the Bay of Bengal. The country's capital is Dacca, which is inland. Fewer than 10 percent of Bangladeshis live in cities, however.

56. Pakistan's capital was Karachi first because it was a Moslem stronghold, and on the sea. The country's real emotional heart was inland, but Lahore was too close to Hindu India. Later, when the Pakistanis gained confidence, they moved to Rawalpindi, an old inland city, while constructing a new capital at Islamabad in the north. The entire north of Pakistan feels under pressure from India, and the Pakistanis were making a bold statement of possession and confidence.

57. The majority of the people in Sri Lanka are Aryans, whose home in ancient times was northern India.

58. Nepal is the world's only Hindu monarchy. This small nation extends from the subtropical lowlands of the north Indian plain to the crest of the Himalayas.

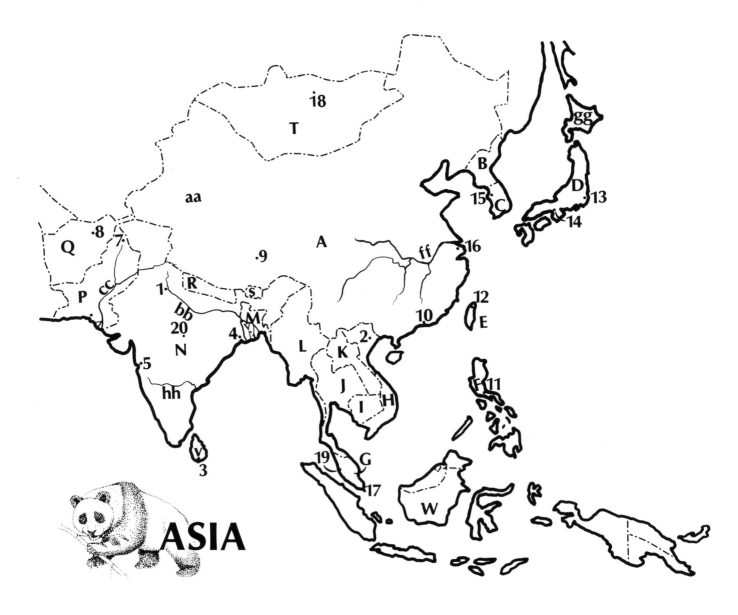

ASIA

ASIA

Locate the following on the accompanying map of Asia (countries are marked with letters; cities with numbers; rivers, mountains, and regions with double letters):

China _____
Japan _____
India _____
New Delhi _____
Nepal _____
Hanoi _____
Hong Kong _____
Thailand _____
Burma _____
Vietnam _____
Manila _____
Taipei _____
Kyoto _____

Seoul _____
Shanghai _____
Bombay _____
Taiwan _____
Mongolia _____
Takla Makan Desert _____
Afghanistan _____
Sri Lanka _____
Bangladesh _____
Singapore _____
Malaysia _____
Ganges River _____
Indus River _____

Chang Kiang (Yangtze River) _____
Calcutta _____
Hokkaidō _____
Laos _____
Bhopal _____
Tokyo _____
Bhutan _____
Ulan Bator _____
Kuala Lumpur _____
Lhasa _____
Krishna River _____

ASIA MAP QUIZ

China **A**	Kyoto **14**	Ganges River **bb**
Japan **D**	Seoul **15**	Indus River **cc**
India **N**	Shanghai **16**	Chang Kiang (Yangtze River) **ff**
New Delhi **1**	Bombay **5**	Calcutta **4**
Nepal **R**	Taiwan **E**	Hokkaidō **gg**
Hanoi **2**	Mongolia **T**	Laos **K**
Hong Kong **10**	Takla Makan Desert **aa**	Bhopal **20**
Thailand **J**	Afghanistan **Q**	Tokyo **13**
Burma **L**	Sri Lanka **V**	Bhutan **S**
Vietnam **H**	Bangladesh **M**	Ulan Bator **18**
Manila **11**	Singapore **17**	Kuala Lumpur **19**
Taipei **12**	Malaysia **G**	Krishna River **hh**

USSR

The USSR is complicated—even more so than most countries. Only half of its people—the Russians—are ethnically similar. The rest are a polyglot "nation" from such disparate places as Estonia and Kazakhstan. But it is still Russians, speaking Russian, who run Russia, from China, to the border of Afghanistan—and, since they invaded that country, beyond. Russians are in charge because the Soviet Union is the result of Russian colonialism, the conquest of other nations nearby. The USSR is perhaps the largest empire ever built by the simple expedient of spreading into ever more distant contiguous states. This created the need to control a lot of rebellious people, which in turn contributes to the suspicious and paranoid style of the Soviet government.

1. The USSR is a large country. In area, would you say it is the world's largest, second largest, or third largest nation?

2. The USSR borders more countries than any other. How many can you name?

3. Where does the word *Russia* come from?

4. These are Soviet republics. Match them with their capital cities:

1	Russia	a	Baku
2	Kazakhstan	b	Kiev
3	Uzbekistan	c	Riga
4	Tadzhikistan	d	Minsk
5	Turkmenistan	e	Moscow
6	Azerbaijan	f	Tashkent
7	Armenia	g	Alma-Ata
8	Georgia	h	Tallinn
9	Ukraine	i	Yerevan
10	Moldavia	j	Tbilisi
11	Belorussia	k	Dushanbe
12	Lithuania	l	Kishinëv
13	Latvia	m	Vilnius
14	Estonia	n	Ashkhabad

5. Which city is "the Venice of the Baltic"?

6. What was Stalin's answer to the demands of Russian Zionist Jews?

7. What do Joseph Stalin and Jimmy Carter have in common?

8. What is World War II called in the Soviet Union?

9. What does the word *Kremlin* mean?

10. What famous "French" modern painter was born in Vitebsk, a small city in Russia, and then went to art school in Saint Petersburg?

11. Where was Russia's first capital?

12. What group is known as the "Little Russians"?

13. A city in the USSR was once second only to Mecca as a holy place of Islam—what city is that?

14. Before the Soviet Revolution in 1917, which country was Russia's chief trading partner?

15. Where was the Charge of the Light Brigade?

16. A person living in the Soviet Union spends an average of twenty-five minutes per day standing in line—for what?

17. The tallest freestanding statue in the world depicts a woman representing a cultural concept that is important to the country she is found in. You've deduced she is not the Statue of Liberty and is probably in the Soviet Union. Very good. What does she represent?

18. We've said that the USSR has little sea access. Yet the USSR has more Pacific Ocean coastline than has the United States—five thousand miles of it. How can both statements be true?

19. What is the buffer state between the USSR and China?

20. From where does the title *tsar* or *czar* derive?

21. We have seen that the Soviet Union is, in terms of its climate, a cold, dry, windy, and generally unpleasant place. This is largely because it is so far north. About what percentage of the USSR is north of the Great Lakes?
30 percent
60 percent
80 percent

22. The United States bought Alaska in 1867 for just over $7 million—from the Russians, who had been exploring and colonizing this area since the mid-1700s. They crossed Siberia and the Bering Strait and explored what is now the U.S. and Canadian coasts. How far south did they come?

23. Which part of the USSR has about 3 percent of the country's land, a fourth of the country's population, and is the leading region in both industrial and agricultural production?

24. With which of the many countries bordering the USSR is it most likely to engage in a serious border dispute?

25. In 1900, the Russian population was twice that of the United States: about 125 million. Yet today the Soviet population is scarcely larger than the U.S. population. Why?

USSR

1. The USSR is twice as big as Canada, which is the second largest country.
2. The Soviet Union shares borders with twelve countries: North Korea, China, Mongolia, Afghanistan, Iran, Turkey, Romania, Hungary, Czechoslovakia, Poland, Finland, and Norway. The United States and Japan are near neighbors across narrow straits.
3. The Rus were Viking traders and pillagers who controlled what is now western Russia more than a thousand years ago. Since today's Russians are Slavs, and the Rus were Germanic Scandinavians, the term *Russia* has annoyed the most chauvinistic of the Slavic Russians ever since.
4.
1–e	8–j
2–g	9–b
3–f	10–l
4–k	11–d
5–h	12–m
6–a	13–c
7–i	14–h
5. Leningrad is the Venice of the Baltic. Like Venice—and Amsterdam and Stockholm—it is built on a network of canals.
6. When Zionist Jews in the Soviet Union demanded a homeland, Stalin's answer was to create a special Soviet Jewish homeland near Birobidzhan, along the Chinese border in far eastern Siberia. Few Jews went.
7. Both were born in Georgia.
8. Soviets know World War II as the Great Patriotic War, in which their losses were extraordinary. More than 20 million Soviet citizens were killed. U.S. losses were about 400,000.
9. *Kremlin* means "castle." Kremlins were the castles or forts of medieval princes in Russia. Many Soviet cities have restored their kremlins, and these now serve as museums or government offices.
10. Marc Chagall was Russian born and trained. He moved to Paris in 1910 at the age of twenty-three.
11. Kiev became the first Russian capital about 1054. In those days the aristocracy were mainly Scandinavians, the lower classes Slavs.
12. The Ukrainians are sometimes known as the Little Russians, although they are of ordinary stature.
13. Bukhara, a city of mosques and minarets in Uzbekistan, Soviet central Asia, was second only to Mecca as a holy place. Besides Communism, Islam may be the most widely observed religion in the Soviet Union.
14. The United States was czarist Russia's main trading partner before the revolution, providing more than 60 percent of its imports.
15. The Charge of the Light Brigade occurred on October 25, 1854, during the Crimean War, near the coastal village of Balaklava, in what is now the USSR. This village also gave its name to a type of warm head covering that leaves only the face exposed.
16. The average Russian stands in line about twenty-five minutes a day to buy food—37 billion hours per year, in the aggregate, or 101 million hours a day, if the stores were open every day.
17. The statue in question is 270 feet tall—versus the Statue of Liberty's runtish 151 feet when measured

sans base—and she represents the "motherland." She stands in Volgograd.

18. Both statements are true, because while the USSR does have a long Pacific coastline, most of it lies very far north and on the wrong—the very coldest—side of the Pacific. So much of the coast is icebound most of the time. Seattle and San Francisco are usable ports all year, but Vladivostok, at a latitude about halfway between them, needs icebreakers to remain open all winter.

19. The buffer between the USSR and China is Mongolia. While Mongolia's historic influences have been mostly Chinese, its recent influences—including the army divisions stationed inside her borders—are Soviet.

20. The word *czar* comes from the Mediterranean region and the word *caesar*.

21. About 80 percent of the Soviet land area is north of our Great Lakes, open to the harsh, frigid air masses of the Arctic. Western Europe gets all the climatic benefits of the Atlantic Ocean, making the USSR even bleaker.

22. The Russians, exploring the U.S. and Canadian coasts, had founded villages as far south as San Francisco Bay. They built a fort—Fort Ross—just north of there in 1812.

23. The Ukraine, in the west, north of the Black Sea, is the Soviet Union's agricultural and industrial anchor. *Ukraine* means "frontier."

24. China disputes its border with the Soviet Union all the time, despite treaties signed in 1858 and 1864. The Chinese claim a large part of eastern Siberia.

25. Many factors suppressed Soviet population growth in the twentieth century. Two world wars and the revolution, civil war, and resulting famine have cost the Soviets at least 70 million lives, if you include children who would have been born if not for these calamities. In all its twentieth-century wars, the United States has lost under 1 million. Simultaneously, immigrants were pouring into the United States—and out of the Soviet Union.

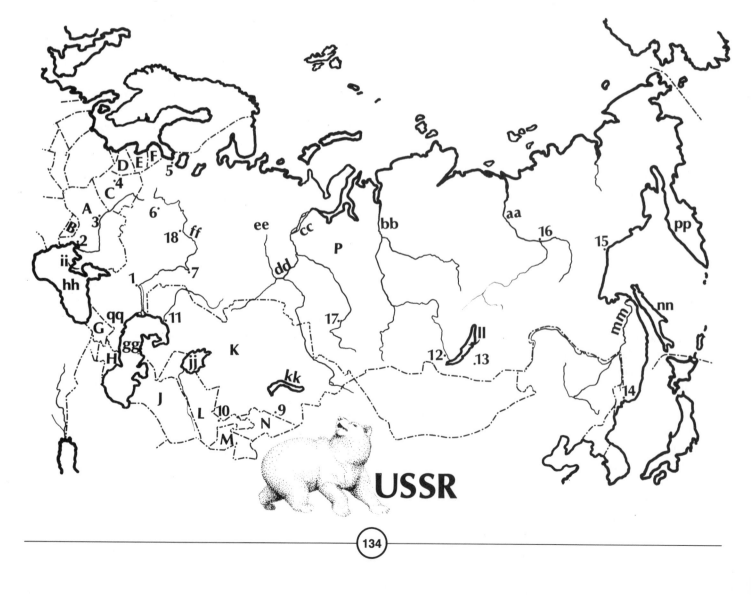

USSR

USSR

Locate the following on the accompanying map of the USSR (Soviet republics are marked with letters; cities with numbers; and rivers, mountains, regions, and so on with double letters):

Ukraine _____
Belorussia _____
Lithuania _____
Latvia _____
Estonia _____
Turkmenistan _____
Uzbekistan _____
Kazakhstan _____
Georgia _____
Azerbaijan _____
Kirghiz _____
Tadzhikistan _____

Moscow _____
Gorki _____
Kiev _____
Kuybyshev _____
Tashkent _____
Alma-Ata _____
Irkutsk _____
Novosibirsk _____
Vladivostok _____
Minsk _____
Leningrad _____
Volgograd _____

Lake Baikal _____
Crimea _____
Ob River _____
Amur River _____
Sakhalin _____
Kamchatka _____
Lena River _____
Ural Mountains _____
Volga River _____
Black Sea _____
Caucasus Mountains _____
Aral Sea _____

USSR

Ukraine **A**	Moscow **6**	Lake Baikal **ll**
Belorussia **C**	Gorki **18**	Crimea **ii**
Lithuania **D**	Kiev **3**	Ob River **cc**
Latvia **E**	Kuybyshev **7**	Amur River **mm**
Estonia **F**	Tashkent **10**	Sakhalin **nn**
Turkmenistan **J**	Alma-Ata **9**	Kamchatka **pp**
Uzbekistan **L**	Irkutsk **12**	Lena River **aa**
Kazakhstan **K**	Novosibirsk **17**	Ural Mountains **ee**
Georgia **G**	Vladivostok **14**	Volga River **ff**
Azerbaijan **H**	Minsk **4**	Black Sea **hh**
Kirghiz **N**	Leningrad **5**	Caucasus **qq**
Tadzhikistan **M**	Volgograd **1**	Aral Sea **jj**

SOUTH AMERICA

South America's population density is relatively low, but is growing dangerously fast; it owns some of the world's richest resources, but they are concentrated in the hands of a very few. It has some of the loveliest scenery in the world, much of it (fortunately, and decreasingly) inaccessible. Its strongest political and emotional ties are not between its countries, but rather to the mother countries and the Catholic Church, which still dominates much of life. It is an enormous continent, but virtually the entire culture of South America derives from one tiny corner of Europe: the Iberian Peninsula.

1. Where are the best Panama hats made?

2. **a** If your friend Ron is a *carioca,* where is he from?
   b Where is a *limeño* from?
   c A *porteño?*
   d A *paceño?*
   e A *paulistano?*

3. What South American country sent troops to fight in Europe during World War II?

4. What country do the Galápagos Islands belong to?

5. Paraguay has an odd sort of comico-sinister reputation, but like any other country it is in business. What is Paraguay's main export?

6. There's something curious about more than half of Paraguay's labor force—what is it?

7. What are the two South American countries with no known oil reserves?

8. In early 1986 only two Latin American countries were under military dictatorships. How many can you name?

9. In Argentina, at the civil registry where births are recorded, is a list of about five hundred first or given names. What is this list used for?

10. Spain's colonies in the New World were divided into four vice-royalties. Describe the approximate present-day area each one controlled:
New Spain
Grenada
New Castile
Rio de la Plata

11. The official name of Uruguay is Oriental Republic of Uruguay. What does the name *Uruguay* mean?

12. Where would you find the Moonies most entrenched in South America?

13. A magazine in Brazil, called *Quac,* has been in publication since 1950. What is *Quac*'s subject?

14. Only two South American capital cities are situated in federal districts, like Washington, D.C. Can you name them?

15. Where was Anastasio Somoza, the late Nicaraguan dictator, assassinated? And how?

16. Where wasn't Mengele hiding?

17. Brazil is an enormous country, so big that only two other countries in South America do not share a border with it. Which two are these?

18. Only two South American countries are landlocked. Which are they? Hint: It is often said that there are retired Nazis living in both.

19. If you go to the *altiplano,* you are almost certain to come down, so to speak, with *soroche.* Please explain.

20. Above the tree line in the Andes is a zone of high-altitude vegetation called the *puna.* The plants here are low shrubs and bunch grasses. What animals might you see grazing upon them?

21. In many Brazilian movies, you will meet characters called *cangaceiros.* What sort of folk are these?

22. Can you guess what product built Brazil's first fortunes? Hint: It wasn't Brazil nuts.

23. People bring twenty thousand live snakes from the Amazonian forests into São Paulo every year. Why, for God's sake?

24. Some countries in South America—Colombia and Venezuela, for example—are primarily *mestizo.* What does that mean, exactly?

25. In certain other sections of South America—northeastern Brazil, for example—there are many *zambos.* What does this mean?

26. If you are very smart and knew the answers to the previous two questions, or cheated and looked them up, you know that the italicized words refer to certain sorts of people. In one part of South America, especially, people keep track of such matters in detail. Here, a *srana* is a person with black skin and red hair; an *escuro* is a mulatto with very dark skin but European features; a *moreno* is a person of typical Mediterranean look; a *cabra* is a light-skinned mulatto; a "Cape Verde" is a mulatto with straight hair; a *chulo* is a person with very light skin and curly hair. Which country is this—one of the most racially integrated in the world, incidentally—where the people engage in this strange, obsessive taxonomy?

27. Cut flowers, pop-up books, clothing, marijuana, and cocaine—what country has a reputation as a great producer of these trade items?

28. Match the Latin American dance with its country of origin:

1 Joropo	**a** Argentina
2 Samba	**b** Venezuela
3 Tango	**c** Brazil

29. Many years ago the wife of Lúcio Costa, one of Brazil's most distinguished architects, was struck and killed by a car. What effect did this personal tragedy have on the workings of Brazil's new capital city, Brasília?

30. Belo Horizonte is a city of 2 million, and the capital of its Brazilian state, a region of general mining—mostly iron ore and coal—called Minas Gerais. What does *mina gerais* mean?

31. Devil's Island, the infamous French prison where Alfred Dreyfus was incarcerated and from which Steve McQueen escaped in *Papillon,* is on the coast of French Guiana, a "colony" of France. For what is French Guiana famous today?

32. Who lives in inland forests of the small country called Suriname?

33. One of the smallest countries in South America is known as the Land of the Six Peoples. This refers to the Amerindians, blacks, and descendants of British, Portuguese, and Chinese settlers, as well as East Indians, who are nearly half the population. Which country is this?

34. If you lived in an Andean Indian village and wanted good luck, what would you bury under your doorstep and in your fields?

35. From where does the Brazil nut come?

36. Who are *los campesinos* and who are *la oligarca?*

37. Tierra del Fuego is way down at the southern tip of South America, and its name means "land of fire." Tierra del Fuego is awesomely cold, with almost constant storms of sleet and blasts of icy wind. It got its name because early European explorers who passed by always saw fires burning on the shore. What did the native Indians of Tierra del Fuego wear?

38. Why is it said that the gauchos, cowboys of the Argentine pampas, eschewed forks?

39. Like any smart tyrant, Hitler banned newspapers that criticized him—even foreign newspapers. The very first one he banned was published in a South American country. Which one?

SOUTH AMERICA

1. The finest grade of Panama hat is called Montecristi and is made in Ecuador, where most Panama hats are made. They came to be called "Panama" hats after the country that imported them and then shipped them on to the United States.

2. **a** A *carioca* such as your friend Ron is a person from Rio de Janiero, Brazil. (*Ron* is Spanish for "rum," and there's a brand of rum called Ron Carioca.)

   b A *limeño* is a person from Lima, Peru.

   c. A *porteño* is a person from Buenos Aires, Argentina.

   d A *paceño* is a person from La Paz, Bolivia.

   e A *paulista* is a person from São Paulo, Brazil.

3. Troops from Brazil joined the Allied campaign in Italy during World War II.

4. Ecuador owns the Galápagos Islands.

5. Paraguay's main export is yerba maté, an herb tea noteworthy as one of only five known natural sources of caffeine.

6. More than half of Paraguay's workers don't work in Paraguay: they migrate to Brazil, Argentina, and even Venezuela.

7. Neither Uruguay nor Paraguay has any known oil.

8. Suriname and Chile were the only Latin American countries under military rule in early 1986. That does not mean that other forms of dictatorship and oligarchical rule were not widespread.

9. The list of five hundred names tells new parents what they are permitted to call their babies when newborns are brought in for registration. Despite a certain liberalization with the return of democracy in 1983, don't try to name your kid Debbie or Scott or Kim or Jennifer—not in Argentina. A name is unacceptable if it is "extravagant, ridiculous, contrary to Argentine customs or ideology or could cause confusion" about the baby's sex. Demostenes and Egipto are fine. But no boys named Sue, for example. No Juice Newtons. No Evelyn Waughs. No Stockard Channings. "People keep mixing up democracy with doing whatever you want," said Dr. Ángel Hernan Lapieza, the registry's director. (Ángel is a fine old Spanish name for a man.)

10. New Spain covered California, Mexico, Central America, and the Caribbean; Grenada covered Venezuela, Colombia, and Ecuador; New Castile covered Peru and Chile; and Rio de la Plata ("Silver River") covered Bolivia, Paraguay, Argentina, and Uruguay.

11. *Uruguay* means "purple land," perhaps for the wildflowers that once covered local hills.

12. The Moonies, followers of the Rev. Sun Myung Moon, and friends seem to be trying to buy Uruguay. They already own the third largest bank, the biggest hotel in Montevideo, the biggest publishing company, a restaurant, a meat-packing company, and other businesses.

13. *Quac* is the magazine that details the adventures of Donald Duck. "Pato Donald" is big in Brazil, where he earns more than $17 million a year from the sale of films, books, and magazines like *Quac*. In 1984, his fiftieth birthday, he had a year-long film festival.

14. Caracas and Brasília are in federal districts.

15. Somoza was killed in Asunción, Paraguay, in September 1980, in a bazooka attack. He is buried in

Miami, where he was fond of spending a lot of his time and his people's money.

16. Mengele wasn't in Paraguay, even though that country gave the notorious Nazi doctor citizenship in 1959 and Nazi-hunters have claimed for years that he was there. But from 1961 to 1975, Mengele lived in Brazil. One reason we didn't know this was that opponents of Paraguay's dictator, General Alfredo Stroessner, used the Mengele claim to help discredit the government.

17. Chile and Ecuador do not share borders with Brazil.

18. Bolivia and Paraguay are landlocked. But Bolivia still claims part of northern Chile, which it lost in the War of the Pacific in 1879–83, and that territory has a coastline.

19. The *altiplano* is the high plain, which averages 13,500 feet above sea level in Bolivia & Peru. *Soroche* is the mountain sickness of the Andes, caused by high altitudes and limited oxygen.

20. Llamas, vicunas, and alpacas would be grazing in the *puna*.

21. *Cangaceiros* are the notorious highwaymen of northeastern Brazil, now romanticized in novel and film, just as Robin Hood and the Godfather's troops are in North America today. These thieves are said to possess honor.

22. Pão Brazil, or brazilwood, from which a once-valuable red dye could be extracted, was the first important product of this region. The wood traders and dye workers were called *brasileiros*, and gave their name to the country.

23. The snakes are collected by the Butantã Institute, founded 1902, which uses them to manufacture antivenom serums. Brazilians need these, because the country has forty-three species of venomous snakes. Delivery of a live poisonous snake gives you credit for some serum to take home whenever you need it. The institute also develops serums from spider and scorpion venom.

24. A *mestizo* is a person of mixed Indian and European ancestry.

25. A *zambo* is a person of mixed Indian and African ancestry. African slaves were brought to the New World to work on tropical-crop plantations, and their descendants, intermixed with Indians and Europeans, are common along the Caribbean coast.

26. The racial taxonomists are the Brazilians, whose citizenship includes all these types and many more.

27. Colombia.

28. 1–b
2–c
3–a

29. When Costa helped to design central Brasília, he saw to it that concrete barriers and tunnels separated the speeding cars from pedestrians. Today, central Brasília (at least) is one of the world's safest cities for walking.

30. The phrase *mina gerais* means "general mining."

31. French Guiana today is the European Space Agency's launching site.

32. The people who live in Suriname's wild inland forests are called Bush Negroes. They are the descendants of escaped slaves and often inhabit villages that resemble West African villages of past centuries. In other parts of Suriname there are Hindus and Indonesians.

33. Guyana, where Jim Jones led his followers and served them grape-flavored Kool Aid, is the Land of the Six Peoples.

34. You'd bury a dried llama embryo everywhere you wanted luck or fertility.

35. You guessed it—the Brazil nut comes mostly from northern Bolivia, where the most abundant stands of *Bertholletia excelsa* can be found. The nut has been exported largely through Brazil, hence its name.

36. *Campesinos* are the Latin American peasants who do the work but own little or nothing, and the *oligarca*—the oligarchy—is the handful of rich, landowning families who rule and reap. This is pretty much the way things work throughout Latin America, and it is why they have civil unrest down there.

37. Nothing.

38. The gauchos said that if you need a fork, then you need a plate, and if you need a plate, you need a table, and then you need a chair, and before long you're living in a house with a woman and a bunch of kids and a dog that kills your chickens. So forget forks. Stick with your *caballo* and eat with your *manos*, say the gauchos.

39. The first overseas paper Hitler banned—in 1933—was the *Argentinisches Tageblatt*, of Buenos Aires. Despite our present-day impression of a 1940s-style European fascism persisting today in South America, many early immigrants from Germany were social democrats and set a rather liberal tone.

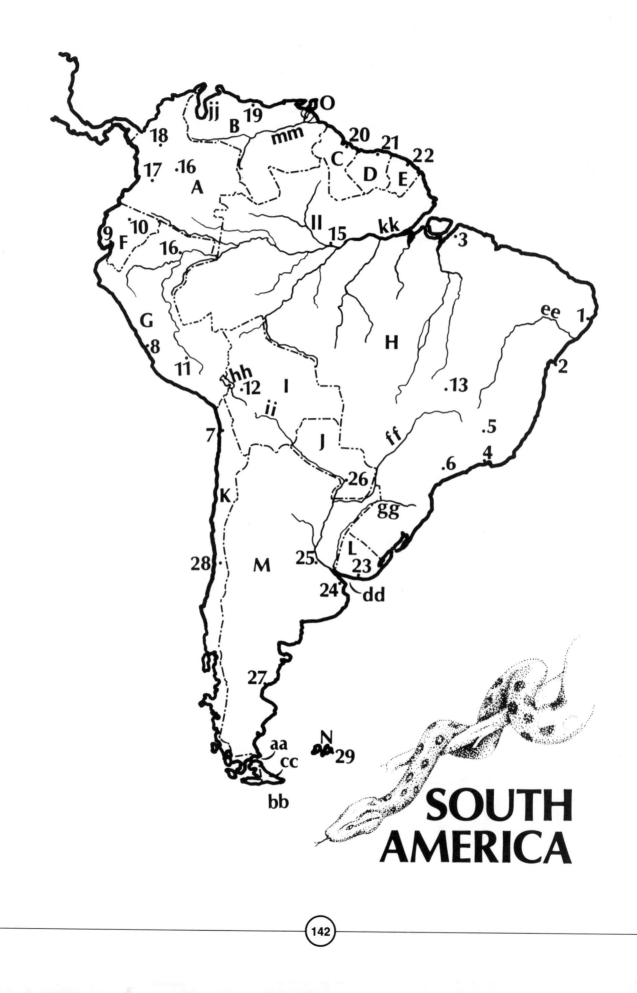

SOUTH
AMERICA

SOUTH AMERICA

Locate the following on the accompanying map of South America (countries are marked with letters; cities with numbers; rivers, mountains, and lakes with double letters):

Brazil _____
Buenos Aires _____
Amazon River _____
Colombia _____
Peru _____
Argentina _____
Rio de Janeiro _____
Chile _____
Venezuela _____
Tierra del Fuego _____
Bogotá _____
Lake Titicaca _____
Ecuador _____
Guyana _____

Suriname _____
Brasília _____
Bolivia _____
Montevideo _____
Paraguay _____
Rio de la Plata _____
Asuncíon _____
Manaus _____
Iquitos _____
Orinoco River _____
Cayenne _____
Quito _____
La Paz _____
São Paulo _____

Recife _____
Uruguay _____
Paraná River _____
Falkland Islands _____
Strait of Magellan _____
Río Negro _____
Lima _____
Caracas _____
Cuzco _____
Stanley _____
Belém _____
Belo Horizonte _____

SOUTH AMERICA MAP QUIZ

Brazil **H**

Buenos Aires **24**

Amazon River **kk**

Colombia **A**

Peru **G**

Argentina **M**

Rio de Janeiro **4**

Chile **K**

Venezuela **B**

Tierra del Fuego **cc**

Bogotá **16**

Lake Titicaca **hh**

Ecuador **F**

Guyana **C**

Suriname **D**

Brasília **13**

Bolivia **I**

Montevideo **23**

Paraguay **J**

Rio de la Plata **dd**

Asuncíon **26**

Manaus **15**

Iquitos **16**

Orinoco River **mm**

Cayenne **22**

Quito **10**

La Paz **12**

São Paulo **6**

Recife **1**

Uruguay **L**

Paraná River **ff**

Falkland Islands **N**

Strait of Magellan **aa**

Río Negro **ll**

Lima **8**

Caracas **19**

Cuzco **11**

Stanley **29**

Belém **3**

Belo Horizonte **5**

CARIBBEAN AMERICA

What we have here is an intercontinental land bridge, part of a continent and an array of islands. The last represent an astonishing range of cultures, from Dutch to Spanish to African. In some places—such as southern Mexico and northern Central America—vestiges of the native Indian cultures have been permitted to survive.

The Caribbean islands are generally an Afro-American realm. Here, the imprint of plantation slave culture has wiped out earlier Indian ways. Middle America is the geographer's name for the Central American isthmus and mainland Mexico, whose culture has been more influenced by Europe.

This region is a particularly clear example of what

geographers call "sequent occupance," or occupation by successive waves of people with different cultures and different ideas of how the land should be used. First were the Indians, then came the Spanish colonists with their Iberian farming/ranching tradition, then the Europeans and North Americans with their labor-intensive, single-crop, factory-type plantations. In the modern period, the plantations are being converted into freehold cooperatives, and the importance of manufacturing is growing. The factories make items for export to the European and North American markets. Today, the entire region is dominated by a more recent outside power, the United States.

1. What is "the Switzerland of Central America"?

2. Which U.S. founding father was born on a Caribbean island?

3. There is a manmade lake, called Gatun Lake, in a very odd place—where?

4. To which country does this sentence from the *National Geographic* refer: "In Central America's most peaceful nation, geography is the only extreme"?

5. In what Central American country are 90 percent of the people of almost pure European descent, with little local Indian blood?

6. What Caribbean country did the United States occupy from 1898 to 1902?

7. Which two Caribbean countries did the United

States occupy from 1912 to 1933 and from 1915 to 1934, respectively?

8. Off the Venezuelan coast are three islands whose official name is the Netherlands Antilles. They are sometimes called the "ABC Islands." What are their individual names?

9. What is *papiamento?*

10. Why is Mérida, on the Yucatán peninsula, called "the white city"?

11. From whom did the United States buy the Virgin Islands?

12. What island in the Caribbean is a "commonwealth," and what country is it part of?

13. As you may know, Tijuana, Mexico, is the birthplace of Caesar salad. How did it (Tijuana, not the salad) get its name?

14. A place called Pointe Salinas has been controversial in recent years. Where is it, and why?

15. In the Caribbean, tourism is often called an "irritant industry." How many ways can you think of that tourists harm these countries?

16. In many coastal cities, especially those of Spanish or Portuguese influence, a common feature is the *malecón.* What is this? How does it differ from a *maricón?*

17. Which Caribbean city became the home of the descendants of Christopher Columbus?

18. In Spain, a *zócalo* is the base of a pedestal. What is a *zócalo* in most of Latin America?

19. All over Latin America, Mexicans suffer ethnic discrimination; they are accused of being slow and dumb. Why should this be?

20. What is the *Rio Bravo del Norte?*

21. Pancho Villa invaded part of the United States from Old Mexico. Which part?

22. The Indians living in Mexico grew three traditional crops. What were these?

23. Besides the ABC Islands, there are three more major islands in what is called the Dutch Caribbean. Can you name them?

24. Peter, Norman, and Jost van Dyke are small islands, members of what colony?

25. As we've mentioned before, sugar both helped found the economies of most Caribbean islands and provides most of their modern revenue. What airport commemorates this fact?

26. Where does reggae music come from?

27. Here are the names of four dead or deposed dictators. Can you tell where each dictated?
Trujillo
Somoza
Batista
Duvalier

28. Which of these is the real "banana republic"?
Costa Rica
Nicaragua
Guatemala
Saint Lucia
Belize
Honduras

29. Only one Caribbean island has major oil reserves. Can you name it?

CARIBBEAN AMERICA

1. Costa Rica is called "the Switzerland of Central America."

2. Alexander Hamilton was born on Nevis, an island in the Lesser Antilles. Sugar was king in the British island colonies of the 1700s, providing wealth far in excess of that provided by the troublesome North American colonies. Many early American business and political leaders, rich men all, visited or came from the sugar islands in the south. George Washington's only travel outside North America was to Barbados as a teenager. Some historians maintain that the American Revolution was won by the decisive activities of allied French warships operating in the Caribbean. Hamilton was the nation's first treasury secretary, and his aristocratic face adorns the ten-dollar bill.

3. Gatun Lake is in the middle of the Panama Canal.

4. Costa Rica.

5. In Costa Rica. This country was settled late by Europeans, because despite its name—meaning "Rich Coast"—it contained no gold. The few Indians who had lived there were already enslaved when Costa Rica's pioneers arrived, and there was little for them to do but farm their own small farms, develop an attitude of independent self-reliance, and marry each other.

6. Cuba.

7. Nicaragua, Haiti.

8. Aruba, Bonaire, Curaçao are the "ABC Islands."

9. *Papiamento* is the language—a mixture of Dutch, Spanish, English, and others—spoken in the Dutch Caribbean. It contains no word for "weather."

10. Mérida, in Mexico, is called "the white city" because its residents once wore mainly white clothes, the streets and sidewalks were perfectly clean, most buildings were whitewashed, and the whole place gleamed in the sun. That was long ago.

11. We bought the Virgin Islands from Denmark in 1917.

12. Puerto Rico, which became a U.S. possession after the Spanish-American War, is a commonwealth of the United States. The interesting thing about this is that few people seem to know exactly what a commonwealth is supposed to be. It is not a state, and not a territory, and not—as the Puerto Rican separatists certainly know—an independent country. What this gives Puerto Rico is all the advantages of close affiliation with the U.S. economy and government aid, but few of the political burdens. What it does not give is the pride of independence.

13. Tijuana, across the border from San Diego, only began growing with the advent of Prohibition in the United States. You could get a drink there and not in San Diego. Before then, this notorious border town was simply *El Rancho de Tía Juana*—or "Aunt Jane's Ranch."

14. Pointe Salinas is the site of the long-runway airport on the island of Grenada. In Washington it was seen as a bad idea when the Cubans were building it. After the United States invaded and occupied Grenada, and had things under control, it was seen as a good idea.

15. Tourism is an irritant industry, and not only in the Caribbean, because tourists are pollution. For all their virtues, and for all the economic benefits they may bring, the free-spending, carefree tourist corrodes native cultures. The local person sees wealth, arrogance, and behavior that is sometimes unwholesome, often supercilious. Tourists demand ever "higher" levels of goods and services—ice, diet soda, expensive imported liquors, shops offering chic clothing—and money from tourists usually goes toward the cost of these imported goods and services. It is never quite enough, so local governments have to pick up the tab for ever splashier airports and smoother highways. Little trickles up, and beyond the tips for service workers, little trickles down.

16. The *malecón* is the oceanfront drive (or, sometimes, riverfront drive) that becomes a public park and promenade. The most famous *malecónes* of the Caribbean are in Havana and Santo Domingo. This raises questions of land use, because this open, public, coastal land use stands in sharp and illuminating contrast to places like Miami Beach, where the public is barred from the sea by private development of exclusive hotels. *Maricón* is the Spanish word for "male homosexual."

17. Santo Domingo, capital of the Dominican Republic, is the home of the descendants of Columbus.

18. In Latin America and the Caribbean, a *zocalo* is the city square or main plaza.

19. There seem to be two explanations, or three if you add the universal human tendency to look down on other peoples. First, Mexican Spanish is spoken slowly and somewhat musically, not unlike English in the southern United States. Secondly, Mexicans have been portrayed in countless Hollywood movies as slow, dumb sidekicks—as comic relief, as Sancho Panzas without Sancho's good sense. These movies then traveled throughout Latin America. If Mexicans are so slow and dumb, how did they develop one of the most successful countries in the New World?

20. *Río Bravo del Norte* is what Mexicans call the Rio Grande.

21. Pancho Villa invaded New Mexico from Old Mexico.

22. The Indians of Mexico grew corn, squash, and beans. They grew them all in the same field, with the squash running between the corn rows and the beans climbing up the stalks. Combining these three plants provides complete nutrition.

23. The ABC Islands are off the coast of Venezuela. In the northern Lesser Antilles are the other three Dutch Caribbean islands: Sint Maarten, Sint Eustatius, and Saba.

24. Peter, Norman, and Jost van Dyke are the smaller islands of the British Virgin Islands. The major island is Tortola.

25. Canefield Airport on Dominica recalls the importance of sugar cane to the Caribbean.

26. Reggae comes from Jamaica. The best sampling of it is found on the sound-track album from the Jimmy Cliff movie, *The Harder They Come*, itself a classic of folk film art. The late Bob Marley, a reggae star, has become a cult hero to many Jamaicans. Reggae comes from the streets of Trenchtown, the Kingston slum, and is philosophically rooted in Rastafarianism, a religion as homegrown as the ganja its adherents smoke.

27. Trujillo dictated in the Dominican Republic, Somoza in Nicaragua, Batista in Cuba, and Duvalier in Haiti.

28. None of the Central American countries mentioned have bananas as their primary crop, if that's what is meant by "banana republic." Most of them grow more coffee and cacao than bananas. But bananas are the primary crop on Saint Lucia, an island in the Lesser Antilles. We never see Saint Lucia's luscious bananas in the United States though, because they all go to Europe.

29. Trinidad, which is geologically linked to nearby Venezuela, is the only island hereabouts with significant oil deposits. It is also the birthplace of one of the greatest living novelists, V. S. Naipaul.

CARIBBEAN AMERICA

Locate the following on the accompanying map of Caribbean America (countries are marked with letters, cities with numbers, geographical features with double letters):

Mexico _____
Puerto Rico _____
Cuba _____
Tijuana _____
Gulf of Mexico _____
Caribbean Sea _____
Panama _____
El Salvador _____
Miami _____
Jamaica _____
Nicaragua _____
Virgin Islands _____
Bahamas _____

Bermuda _____
Mexico City _____
Gulf of California _____
Havana _____
Dominican Republic _____
Costa Rica _____
Guatemala City _____
Curaçao _____
Trinidad _____
Barbados _____
Guantánamo Bay _____
Tegucigalpa _____
Nassau _____

Acapulco _____
Key West _____
Yucatán _____
Haiti _____
Montego Bay _____
Cabo San Lucas _____
Cozumel _____
Mazatlán _____
Grand Cayman _____
Martinique _____
Grenada _____
Isthmus of Tehuantepec _____
Isle of Youth _____

CARIBBEAN AMERICA MAP QUIZ

Mexico **A**	Bermuda **N**	Acapulco **18**
Puerto Rico **M**	Mexico City **19**	Key West **37**
Cuba **I**	Gulf of California **ii**	Yucatán **jj**
Tijuana **25**	Havana **4**	Haiti **K**
Gulf of Mexico **bb**	Dominican Republic **L**	Montego Bay **38**
Caribbean Sea **cc**	Costa Rica **F**	Cabo San Lucas **26**
Panama **G**	Guatemala City **16**	Cozumel **17**
El Salvador **D**	Curaçao **W**	Mazatlán **21**
Miami **1**	Trinidad **V**	Grand Cayman **Z**
Jamaica **J**	Barbados **U**	Martinique **R**
Nicaragua **E**	Guantánamo Bay **5**	Grenada **T**
Virgin Islands **O**	Tegucigalpa **14**	Isthmus of Tehuantepec **ll**
Bahamas **H**	Nassau **3**	Isle of Youth **aa**

NORTH AFRICA AND THE MIDEAST

The most visible characteristic of this region is aridity: It is dry almost everywhere. People have settled along rivers and coasts, where they found some moisture. The second most visible characteristic is discord, strife, and war. There is irony in that, because the Mideast is the birthplace of Judaism, Christianity, and Islam—three of the world's most influential religions. On the other hand, there may be no irony at all, since much of the strife grows out of the very fact that these religions exist here, so close together, and are so powerfully felt by their adherents.

As we shall see, however, there is more to this region than dryness and disharmony.

1. Where is the Sphinx's beard?

2. What country pays its people, per capita, the most in farm subsidies?

3. Why has damming the Nile with the Aswān High Dam been a mixed blessing for Egypt?

4. What do Armenians call their homeland?

5. Why can't you accurately call North Africa and the Mideast the Muslim World?

6. What is the name of the union of several small principalities along the eastern side of the Arabian peninsula, on the Persian Gulf? And what do they mainly produce?

7. How many of the thirteen OPEC countries can you name? How many are found in this region of the world?

8. Where, exactly, was ancient Mesopotamia?

9. Why did such a successful civilization as Mesopotamia not survive into the present?

10. Name the country in the Mideast that is almost as much Christian as it is Muslim. How did it get to be Christian, in the middle of this Islamic region?

11. Why has the incredible wealth brought to many Mideast countries by oil not resulted in a better life-style for more citizens?

12. What country, still a kingdom, occupies the northwest "corner" of Africa?

13. The Arab League was formed in 1945 to promote Arab causes and strengthen ties within the Arab world. Which two countries are not members?
Morocco   Iraq
Algeria   Jordan
Egypt
Oman
Qatar
Iran

14. What event of 1956 first established Egypt as a leader among the Arab countries?

15. When you think of the eastern part of North Africa, around Egypt, the physical feature you remember instantly is a river, the Nile. When you think of the western part of North Africa, another kind of physical feature comes to mind: mountains. Can you name this mountain range?

16. Who are Maghrib's oldest residents? Hint: They gave this area one of its other names, the Barbary States.

17. What does *Zionism* mean?

18. There are many issues that separate the Arabs from the Israelis and have kept them in a virtual state of war for forty years. What would you say is the most difficult of these?

19. Most of Israel has a different look from the rest of the Mideast: more prosperous, more successful, more like a Western, developed country. Seventy percent of Israelis, for instance, live in cities. Why, in all the Mideast, does this seem true only of Israel?

20. Which indisputedly Islamic and Middle Eastern country has remained aloof from the doings of the other Middle Eastern countries since World War I?

21. What was the last name of the Turk who is called the father of modern Turkey?

NORTH AFRICA AND THE MIDEAST

1. The Sphinx lost its three-foot beard in 1825, when antiquarians carted it off to the British Museum. The beard became one of the numberless antiquities that the imperialist countries have stolen from their subject states. Egypt would like it back, just as Mexico and Guatemala would like to have back Mayan antiquities now reposing in the Smithsonian and South Africa would like the diamond known as the Star of South Africa back from the Tower of London, where it is considered to be part of the British Crown Jewels. There has been talk of restoring the Sphinx's face, but the beard is where it is, and the history of such cultural plunder—however it can be justified in some instances—reveals few instances of restitution.

2. Saudi Arabia gives its farmers more subsidies than any of the Western nations. It has to, if it wants to farm at all, because Saudi agriculture is like growing oranges in greenhouses in Siberia—you can do it, but it's not cheap. Most of Saudi Arabia is desert. But in the not-so-dry western part of the country, petrodollars are financing authentic agriculture, and Saudi Arabia is the Mideast's third largest wheat exporter. Saudi Arabia, however, is still the world's third largest food importer.

3. The dam at Aswān has been a blessing, as predicted, regularizing the Nile's floods, permitting year-round irrigation and three crops annually. Today, the Egyptian nation seems to control the Nile, rather than vice versa. But the former flooding provided natural silts, and the fertility they brought is now lost and must be replaced with inorganic fertilizers. Once you begin using them, you have to do it forever. In addition, excess applications of water have led to salinization of the soil, poisoning the crops. And without annual deposits of flood silts, the Nile delta has begun eroding away. And the coastal sardine fishery has collapsed. There is no free lunch.

4. Hayastan. Armenians refer to themselves as Hayq.

5. You can't limit the World of Islam to this area, because Islam extends far beyond it. And, inside this territory, there are regions where Islam is scarcely practiced.

6. The United Arab Emirates, along the Persian Gulf, is a union of seven emirates: Ajman, Abu Dhabi, Dubai, Sharjah, Umm al-Qaiwain, Ras al-Khaimah, and Fujairah. This region is only 32,000 square miles, and the population is only half a million, but it produces about $15,000 worth of oil per person per year.

7. The thirteen members of OPEC are Algeria, Iran, Kuwait, Qatar, Saudi Arabia, United Arab Emirates, Libya, Iraq, Ecuador, Gabon, Indonesia, Nigeria, and Venezuela. The last five are elsewhere.

8. Mesopotamia lay between the Euphrates and Tigris rivers, in what is now Iraq and Syria. There were cities in this area as long as five thousand years ago, and our civilization is built on the creations of their inhabitants.

9. There are many explanations. Invasions and corrupt politics certainly helped. However, some geographers believe changes in environment destroyed Mesopotamia, as they may have altered or destroyed other ancient civilizations. Overgrazing and salinization of irrigated soils led to desertification.

Areas surrounding the cities slowly became desolate.

10. Lebanon is about half Christian, because the spread of Islam in the Middle Ages so threatened the Roman Christians that Crusaders were sent eastward to conquer—and convert. They did so in the eleventh century and left pockets of converts behind. Their descendants are the Lebanese Christians of today.

11. The region's oil wealth has changed the lives of many, especially in the smaller countries of the Mideast, such as Kuwait, which has the world's highest per-capita income. It has gone into such quality-of-life items as housing and health care. But in the larger countries, it is harder to make an impact because of traditional inertia and too many people. There, the people are further away from the government, and the government often either does not care or does not have the stability to act. Apart from Iran, the oil-rich countries of the Middle East have low populations.

12. Morocco is in the northwest corner of Africa.

13. Iran is not a member of the Arab League, because Iranians are not Arabs. They are, for lack of a better modern term, still Persians. As Persia, Iran was a kingdom 2,500 years ago. Nor do Iranians adhere to the Sunni Muslim faith, which is typical in North Africa and the Middle East. Instead, they are Shi'ite Muslims, who amount to perhaps a tenth of all Muslims. Shi'ites are fundamentalists, as we know from the pronouncements of Khomeini and the actions of post-shah Iranians. Egypt was expelled because it made peace with Israel.

14. In July 1956, Egypt nationalized the Suez Canal and took over the firm that controlled it. In response, Israel, France, and England invaded the Sinai Peninsula. But the Egyptians held them off, and pressured by world public opinion (as well as U.S. support for Egypt) and their own people, the three invaders retreated. To the Third World in general, and to other Arabs in particular, Nasser's Egypt became a heroic nation.

15. The Atlas Mountains cover much of inland northwest Africa, in the region called the Maghrib: Algeria, Tunisia, Morocco, and Libya. These mountains catch rainfall, and thus parts of the region are much richer than the area farther east. Rainfall, for instance, is three times higher along the coast of Tunisia than in Alexandria. Where the Atlas Mountains stop, the desert starts.

16. The Berbers, a nomadic hunting people, are the oldest residents of the Maghrib. When the Arabs came and conquered, most Berbers were converted and joined up.

17. Zionism is the aspiration of European Jews for a homeland in the Middle East, and nothing more. But pursuit of this aspiration's fulfillment, and resistance to its fulfillment, are central to much of the Mideast's troubles.

18. Probably the most difficult issue between the Arabs and the Israelis is the question of refugees—the 750,000 Palestinian Arabs who fled their homeland when Israel was created. They, and now their descendants in Egypt, Jordan, Lebanon, and Syria, call themselves "a nation without a state." Which is just what the Jews were before the creation of Israel.

19. Israel's success is based on the enormous intelligence and technological skill of its immigrants, who brought these skills from their former countries in Europe and the West; on their enthusiasm and commitment to building and defending a new homeland; and on the great sums of money donated for this purpose by sympathetic supporters abroad.

20. Turkey. Beginning in the 1920s it removed Islam as official state religion, replaced Moslem law with a modified version of Western law, began the emancipation of women, and generally began to surrender the historic xenophobia of the Arab world and orient itself toward Europe.

21. Ataturk. He was the original "young Turk."

NORTH AFRICA AND THE MIDEAST

Locate the following on the accompanying map of North Africa (countries are marked with letters; cities with numbers; and rivers, deserts, and mountains with double letters):

Algeria _____
Egypt _____
Iran _____
Saudi Arabia _____
Turkey _____
Iraq _____
Tehran _____
Cairo _____
Nile River _____
Casablanca _____
Algiers _____

Ankara _____
Morocco _____
Tunisia _____
North Yemen _____
Oman _____
Niger _____
Mauritania _____
Lebanon _____
Syria _____
Israel _____
Tripoli _____

Riyadh _____
Istanbul _____
Ahaggar Mountains _____
Atlas Mountains _____
Cyrenaica _____
Arabian Desert _____
Aïr Mountains _____
Tigris River _____
Euphrates River _____
Mecca _____
Qatar _____

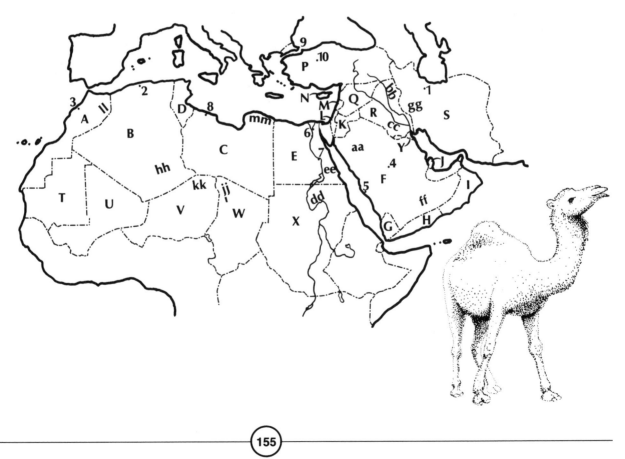

NORTH AFRICA AND THE MIDEAST MAP QUIZ

Algeria **B**
Egypt **E**
Iran **S**
Saudi Arabia **F**
Turkey **P**
Iraq **R**
Tehran **1**
Cairo **6**
Nile River **dd**
Casablanca **3**
Algiers **2**

Ankara **10**
Morocco **A**
Tunisia **D**
North Yemen **G**
Oman **I**
Niger **V**
Mauritania **T**
Lebanon **M**
Syria **Q**
Israel **L**
Tripoli **8**

Riyadh **4**
Istanbul **9**
Ahaggar Mountains **hh**
Atlas Mountains **ll**
Cyrenaica **mm**
Arabian Desert **ee**
Aïr Mountains **kk**
Tigris River **bb**
Euphrates River **cc**
Mecca **5**
Qatar **J**

SUB-SAHARAN AFRICA

Most of the people in Sub-Saharan Africa still live in villages and farm for a living. More are moving to cities daily, but Africa remains the least urban continent. It also has the poorest soil sand the highest birthrate. And in many areas where the soil remains fertile, people use it not to grow food but to produce cash crops for sale abroad, because many governments are more interested in profit than in feeding people. If you read newspapers, you know what sort of trouble these facts add up to. We see one version in Uganda and another, with black-white racism mixed in, in South Africa. These troubles grow out of history and geography, of course. In places like these, geography is there to starve in, fight over, or run to.

1. Where is
   a Tombouctu?
   b Khartoum?

2. What country in Africa has the most people?

3. What country in Africa is the largest in area?

4. Name two of the four countries that border Lake Chad.

5. Match the tribe or ethnic group with the country most of it lives in:

1 Ashanti	**a** Sudan	**9** Kru	**i** Ghana
2 Afrikaner	**b** Kenya	**10** Masai	**j** Madagascar
3 Baggara	**c** Liberia	**11** Merina	**k** Nigeria
4 Dinka	**d** Botswana	**12** Ndebele	**l** Zimbabwe
5 Falasha	**e** Senegal	**13** Tswana	
6 Fellahin	**f** South Africa	**14** Wolof	
7 Ibo	**g** Egypt	**15** Zulu	
8 Kikuyu	**h** Ethiopia		

6. Name the two great rivers of Africa that join to become the Nile—and where do they join?

7. What country is the "Switzerland of Africa"?

8. Three regions of present-day Africa provided most of the slaves to the New World. Where are they?

9. In many African countries—including Botswana, Lesotho, Swaziland, and the Central African Republic—many men have left their country to find work elsewhere. What has this meant to the work force at home?

10. In what country did Albert Schweitzer build his hospital?

11. In many areas of the world, food production is rising due to improved methods of agriculture. But in _____ food production per capita has fallen.

12. Name some ways in which Pygmies are different from most other people.

13. Which African country did Winston Churchill term the "pearl of the Nile"?

14. Which African country had its colonial capital outside its borders?

15. Which four African countries were founding members of the United Nations? Hint: This is the same as asking which were independent in 1945, the year the U.N. was founded.

16. Which African country is the continent's major exporter of oil?

17. Which part of the Sudan would like to be another country?

18. Which parts of Ethiopia would rather be a separate country?

19. The most media-visible characteristic of the Idi Amin regime in Uganda, between 1971 and 1979, was the terrorizing and murder of Ugandan citizens by representatives of their government. At last, the Amin regime was overturned, and Milton Obote returned to the presidency in 1980. About how many Ugandan citizens were killed in political strife between 1980 and 1984?
  a A handful, if any
  b Possibly a few hundred
  c About 100,000

20. Which of these African countries has the greatest percentage of its people in exile?
Guinea
Nigeria
Kenya
Ethiopia

21. What country's motto is "The Love of Liberty Brought Us Here"?

22. The general greeting in Botswana is *pula*. What does this mean? Hint: Most of Botswana's territory is in the Kalahari Desert.

23. Many of black Africa's countries are now named from their indigenous languages: Nigeria, Sierra Leone, and so on. Right?

24. What city was the capital of the British South Africa Company?

25. What are the "rift valleys" of Africa?

26. What characteristic of Africa vis-à-vis South America was noticed so early by explorers that Francis Bacon mentioned it in the 1620s?

27. Where is Gondwana Land?

28. Why are there so few mountains in Africa? (Hint: It has to do with plate tectonics.)

29. Black Africa (another term for Africa south of the Sahara) contains about 8 percent of the world's population (360 million people or more). But it has an even more impressive share of something else—a full third of the world's _____. Fill in the blank.

30. Africa has a rich heritage of folklore, poetry, art objects, buildings, law codes, and habits of social behavior, and Africa may have been the very cradle of humankind itself. Yet among the ignorant, the cultures of Black Africa are sometimes denigrated as impoverished or uncreative. Can you explain this?

31. Here is Africa, in the tropics, with an ocean on each side, and yet there is drought across much of the continent. How can this be?

32. Here is Africa, colonized centuries ago by the most advanced civilizations of Europe, and yet most of its people are still subsistence farmers in remote villages. Why?

33. There is a single tiny creature—some people even joke about it—that prevents vast areas of Africa from being usable as cattle range. What creature is this?

34. Salt and ivory. Explain what these products had to do with the trading patterns of western Africa before the Europeans arrived.

35. What caused the west African savannah cities like Tombouctu to lose influence to newer coastal cities? Hint: Think about what was being traded.

36. If Africans themselves owned slaves before Europeans got into the slave trade—and they did—what was so terrible about the Europeans doing it too?

37. When the Europeans colonized Africa, each country developed a different style of rule. Match the country with its method:
1 Britain   a Paternalism
2 Belgium   b Indirect rule
3 France   c Exploitation
4 Portugal   d Assimilation

38. Why are the countries of Africa that you see on maps shaped the way they are?

39. Can you name the literary-philosophical movement that focuses on the virtues of blackness and on African cultural history and that was born not in Africa but rather in France and the Caribbean?

40. What is a "periodic market"?

41. Kenya and Tanzania are adjacent countries on the Indian Ocean. One has its population, most of whom are subsistence farmers, spread throughout the territory; is socialistic; has many different ethnic groups; has integrated them all into the country's economic and social life; and has managed to attain relative political stability. The other has much of its population concentrated in a core area that is relatively prosperous; is capitalistic; has a larger gross national product than the first country; is largely controlled by one ethnic group, the Kikuyu; and has great gaps between economic classes. Can you tell which is which?

42. In Nairobi, who are the *Wa-Benzi?*

43. A single country in southern Africa has more white inhabitants than all other countries in Black Africa combined. Which country is this?

44. In terms of resources, what is the richest country in all Sub-Saharan Africa?

45. In southern Africa, what country is entirely within the borders of another country?

46. In South Africa, what is the official name of apartheid?

SUB-SAHARAN AFRICA

1. **a** Tombouctu is in Mali.
   b Khartoum is in the Sudan.
2. Nigeria, with about 85 million people.
3. Sudan.
4. One of the countries bordering Lake Chad is Chad; the other three are Nigeria, Niger, and Cameroon.
5.
1–i	9–c
2–f	10–b
3–a	11–j
4–a	12–l
5–h	13–d/f
6–g	14–e
7–k	15–f
8–b	

6. The Blue Nile and the White Nile join at Khartoum, in the Sudan, to form the fabled Nile. The Blue Nile begins in Lake Tana, in the Ethiopian highlands. The White Nile tumbles out of Lake Victoria in Uganda, and nearly dries up in the great swampland of southern Sudan, the Sudd, before it joins the Blue Nile.
7. The Switzerland of Africa is Swaziland.
8. The Guinea Coast, particularly regions of modern Nigeria, Ghana, and Ivory Coast provided many slaves to North America. Mozambique and the Congo-Angola Coast shipped people chiefly to Brazil.
9. In these countries, there are more women wage earners than men.

10. Schweitzer built his hospital in present-day Gabon, on the Atlantic coast of Africa. Today Gabon is third only to Namibia and South Africa in the number of hospital beds, with more than 600 per 100,000 persons—a rate comparable to that of western Europe.

11. In the early 1980s, in all of Black Africa between the Mediterranean countries and South Africa, food production per capita was only 80 percent of what it was in the early 1960s.

12. The Mbuti Pygmies of Zaire's Ituri Forest are smaller than other people, to be sure, with men averaging four feet eight inches and women four feet six inches. They are at least as "primitive" as any other group of people on earth. They have the fewest lines and ridges in their fingerprints. Anthropologists do not know what this means, if anything. By the time an Mbuti woman is twenty-five, she will have walked, barefoot and heavily loaded, a distance equal to the circumference of the earth.

13. Uganda, a country of beautiful green hills, but murderous modern politics.

14. As the Bechuanaland Protectorate, Botswana was administered by the British from Mafeking in South Africa. A new capital, Gaborone, was built in 1966 when Botswana became independent.

15. Egypt, Ethiopia, Liberia, and South Africa were UN members from the start.

16. Nigeria exports more oil than any other African country. It has large deposits beneath the Niger River delta.

17. The southern region of the Sudan would like to be independent. The people here are quite unlike the Islamic, Arabic majority of the country. They are black, Christian or animist, and not adapted—or interested in becoming adapted—to the desert life of the rest of the Sudan.

18. The provinces of Tigre and Eritrea in the north of Ethiopia would love to be separate. These are Islamic, Saharan lands, quite unlike the Amharic, humid highlands of the country's central region, which surrounds the capital.

19. The answer is *c;* at least 100,000 Ugandan civilians have been killed since Amin's downfall. Obote is gone again, and a new military government is trying to sort things out, again.

20. This is hard to answer, because nearly every African country has exiles, and many of them harbor exiles. Political refugees seem to be everywhere. But the probable winner (or loser) is Guinea, a former French colony in West Africa where, during the reign of Sekou Toure (1958–85), about 1.5 million or fully one-fifth of the population left. Toure operated some of the world's worst political prisons.

21. Liberia was founded in western Africa in 1822 by freed American slaves. Slave descendants became an elite aristocracy who ruled the country, with the help of American business interests, for more than 150 years.

22. *Pula* means "rain," and any greeting is a fervent wish for some. The Botswana government has named its unit of currency the *pula*.

23. Some African nations now carry true African names. But most of the names are recent corruptions of colonizers' names, which in turn were corruptions of African words: Zambia, after the Zambezi River; Nigeria, after the Niger River; Namibia, after the Namib Desert. Sierra Leone is European also. *Serra lyoa* was the name Portuguese navigators gave to the thunderstorms that roared out of the region's low mountains.

24. Salisbury, Rhodesia, was the capital of the British South Africa Company. It is now called Harare, Zimbabwe.

25. The rift valleys are huge trenches, usually running north-south, that formed when the earth's crust cracked and portions of it sank to become valleys. They run from Swaziland in the south to the far northern end of the Red Sea. Some, like Lake Nyasa, are filled with water.

26. Bacon, and others, noticed very early that the shape of Africa's west coast looked very much as if it would match the shape of South America's east coast if the two were pushed together—or as if they had once been together. This fact, coupled with many others (including the existence of the rift valleys), helped scientists develop the theory of plate tectonics.

27. Gondwana Land no longer exists. It was not a country, but a supercontinent (also known as Gondwana) that is thought to have existed, with Africa as its central core, before the southern continents as we know them began to drift apart. South America moved away to the west and south, India moved to the northeast, Australia to the east. Africa itself apparently moved a little to the north from its initial position nearer Antarctica. This movement appears to be continuing.

28. Africa indeed has few mountains, and those few it has are often individual volcanic ones, rather than great buckled ranges. The reason, once again, is the

drifting continents. As the other continents moved, they folded and buckled upward, the way a towel would if you tried to push it along a carpet. This made mountains. Since Africa was the place from which other continents drifted away, it has few mountains.

29. Black Africa contains fully a third of the countries of the world.

30. Besides racism, the main reason many people are ignorant of Black Africa's cultural richness and contributions is that Africa has an oral tradition, rather than a written one, so that beyond a certain point, little written history exists.

31. Africa is often dry because rainfall in most places is not great. Rainfall in much of the continent is concentrated in short seasons with long dry spells in between. The heat evaporates much of the rain that does fall so that it cannot be used by plants. Great tropical forests are widespread only in the Zaire basin.

32. Most Africans are still subsistence farmers in villages. This is because the colonizers never intended to modernize Africa, only to exploit its resources. Life for most Africans who were not enslaved or brought to work for Europeans went on unchanged.

33. The tsetse fly infests great areas of tropical Africa, preventing them from being used to grow cattle. The fly is dangerous to humans as well, because it carries African sleeping sickness.

34. Salt is plentiful in the north of Africa, but farther south climates are too humid for deposits to form. Elephants lived in the south, but not in the arid north. People from these different environments traded commodities back and forth, also trading ideas and establishing cultural links. People of the savannah areas in between created entire cities to facilitate this trade. These cities were hubs of commerce and education. But times have changed, and the name of one such city is today a symbol of remoteness and isolation: Tombouctu.

35. The sort of trading that Tombouctu existed to facilitate began to die out when a more profitable trading began: the trade in slaves. For this, coastal cities arose, and the interior declined.

36. The Africans owned slaves differently. For one thing, only a few Africans were rich enough to have them. They treated them well, kept slave families together, permitted marriage, generally provided decent quarters, and made them part of the family. Furthermore, the total number of slaves was very small. What the Arabs and Africans did in collaboration with the whites was to kidnap millions of Africans—perhaps as many as 30 million—allow many of them to die miserably, and then mistreat virtually all the rest.

37. The British method of ruling its African colonies was *indirect rule*. They left the native power structure relatively intact and simply coopted it, ruling through puppet chieftains, who collaborated with the Crown. The Belgians were *paternalistic*, tutoring Africans in European ways as if they were very slow children who were not expected to catch up ever. The French wanted the Africans to *become* French, making the colonies part of overseas France and cramming them with French cultural values and the French way of life. The Portuguese simply *exploited* the African territories, not even pretending to give anything back.

38. The size, shape, and composition of today's African countries have very little relation to the cultures or geography of that continent. They were created by Europeans for their convenience, and after a lot of haggling during which no African was represented, at a conference in Berlin in 1884. The Europeans simply chopped Africa into pieces and handed it out, splitting tribal territories the way the slavers had split families and with as little regard for the long-term effects of their greed and insensitivity. We see these effects today. Or, as one geographer, Harm J. de Blij, puts it, "The African politico-geographical map is a permanent liability resulting from three months of ignorant, greedy acquisitiveness during Europe's insatiable search for minerals and markets."

39. Negritude is the name of the literary-philosophical movement that began among black scholars and writers abroad. One of its leaders, Leopold Senghor, eventually became president of Senegal.

40. A periodic market is a market that takes place at regular intervals—say every three or nine days—in different places in rotation. This African institution is a kind of floating flea market. It serves also as social center and grapevine. Moving the market periodically allows different neighborhoods to gain the profits and benefits of being its host.

41. The first country is Tanzania and the second is Kenya.

42. The *Wa-Benzi*, in Nairobi, are that class of people who own and drive expensive automobiles.

43. South Africa has more whites than all other Black African countries combined. But whites are still vastly in the minority there. Of its 29 million people, 21

million are Africans and only 4.5 million are white. The rest are "Coloureds" (2.5 million, of mixed white and African ancestry) and Asians (1 million, mostly of Indian ancestry).

44. South Africa is the richest country in Sub-Saharan Africa—by far.

45. The country of Lesotho is entirely within the country of South Africa. But while South Africa is very rich, mountainous Lesotho is very poor.

46. The official name of apartheid is "separate development."

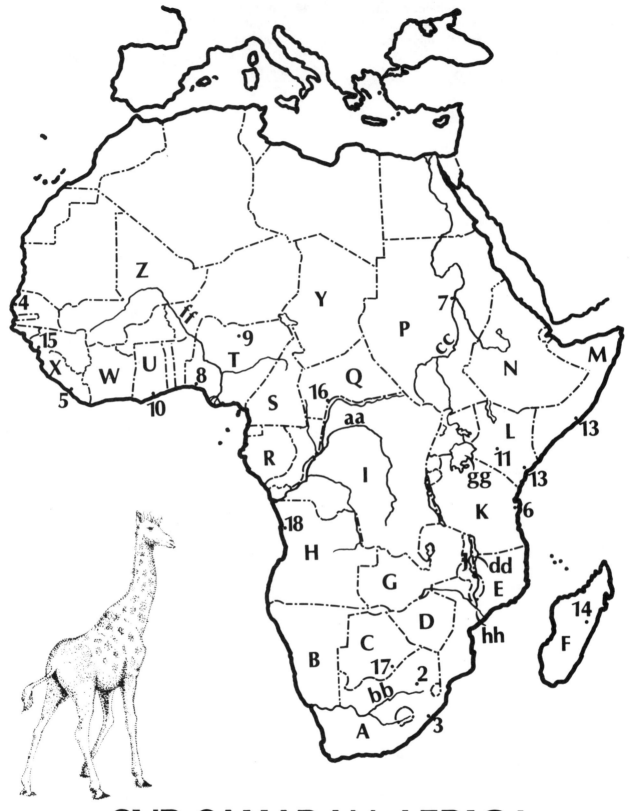

SUB-SAHARAN AFRICA

SUB-SAHARAN AFRICA

Locate the following on the accompanying map of Sub-Saharan Africa (countries are marked with letters; cities with numbers; rivers, lakes, and mountains with double letters):

South Africa _____

Kenya _____

Zaire _____

Madagascar _____

Congo/Zaire River _____

Angola _____

Zambia _____

Nile River _____

Nigeria _____

Ethiopia _____

Sudan _____

Gabon _____

Dakar _____

Lake Nyasa _____

Chad _____

Niger River _____

Lake Victoria _____

Ivory Coast _____

Monrovia _____

Johannesburg _____

Tanzania _____

Somalia _____

Dar es Salaam _____

Zimbabwe _____

Central African Republic _____

Zambezi River _____

Vaal River _____

Khartoum _____

Mozambique _____

Antananarivo _____

Kano _____

Mali _____

Namibia _____

Conakry _____

Bangui _____

Gaborone _____

Nairobi _____

Accra _____

Luanda _____

SUB-SAHARAN AFRICA MAP QUIZ

South Africa **A**	Lake Nyasa **dd**	Vaal River **bb**
Kenya **L**	Chad **Y**	Khartoum **7**
Zaire **I**	Niger River **ff**	Mozambique **E**
Madagascar **F**	Lake Victoria **gg**	Antananarivo **14**
Congo/Zaire River **aa**	Ivory Coast **W**	Kano **9**
Angola **H**	Monrovia **5**	Mali **Z**
Zambia **G**	Johannesburg **2**	Namibia **B**
Nile River **cc**	Tanzania **K**	Conakry **15**
Nigeria **T**	Somalia **M**	Bangui **16**
Ethiopia **N**	Dar es Salaam **6**	Gaborone **17**
Sudan **P**	Zimbabwe **D**	Nairobi **11**
Gabon **R**	Central African Republic **Q**	Accra **10**
Dakar **4**	Zambezi River **hh**	Luanda **18**

AUSTRALIA AND THE SOUTH PACIFIC

Australia has a lot of land (3 million square miles) and not many people (about 4.5 per square mile). Most of the land is arid; only 1 percent is fit for cultivation. Yet the people are known everywhere as fun-loving and bright, and they enjoy one of the highest standards of living in the world. They produce enormous agricultural surpluses to export. They are also lucky. Australia has vast natural resources, much of them yet undeveloped.

New Zealand is just like Australia, some might say, only smaller, quieter, and greener, and its most ambitious young people often move to Australia.

1. What is a three-dog night?

2. Which is not one of Australia's seven states: Northern Territory, Western Australia, Queensland, New South Wales, Victoria, New Brunswick, South Australia, Tasmania.

3. What is the Australian national anthem? Hint: You'll be surprised to learn you can already sing it.

4. Where do most people in Australia live?

5. Some new names have appeared in the South Pacific recently. Here are three:
   a Kiribati
   b Vanuatu
   c Tuvalu
Can you give their previous names?

6. New Zealand is composed of two large islands—North Island and South Island—and many smaller islands. What is south of North Island?

7. What would you say is the sheep-to-person ratio on New Zealand?
One to one   Twenty to one
Ten to one   One hundred to one

8. The world's third largest concentration of ethnic Greeks is a city in Australia. Which city is it?

9. The Australian state of Victoria has the same nickname as the U.S. state of New Jersey. Do you know what it is?

10. In *Gulliver's Travels,* Jonathan Swift locates certain odd characters in Australia. They were brutish, degraded creatures called Y_____s. Fill in the blank.

11. Translate the following into standard English from Australian slang:
   a *To barrack for*
   b *Bluey*
   c *Bonzer*
   d *On the outer*
   e *Tucker*
   f *Digger*

12. More than 20 percent of all Australians live in or near the same city. Which one?

13. What ethnic group in Australia is almost entirely excluded from the modern society?

14. If you lived in eighteenth-century Britain, could not pay your debts, came before a judge, and were sentenced to "transportation"—what would become of you?

15. Perth is on the far western edge of Australia. Sydney is on the far eastern edge. They are about as far apart as the distance from Los Angeles to:
Jacksonville
Atlanta
Chicago
Denver
Hollywood

16. Australia is roughly the same area as the United States (less Alaska). But its population is less—about the same as the U.S. population was in the year:

1776	1898
1805	1917
1825	1944

Hint: That population is 15 million.

17. Sydney and Melbourne were always quarreling about something, so it's no wonder neither could become the country's capital. And Perth was too far away. So, what is the capital of Australia?

18. For years, immigration into Australia was forced, or had to be subsidized. But in 1851, something changed that. What was this?

19. If you've heard of any of Australia's mining regions, it is probably Broken Hill. What is mined there?

20. Australia has more than one _____ (fill in the blank) for every three people. Hint: This is needed to get around in such a large, empty country.

21. The labor movement in Australia, until recently, has thrown its weight behind a certain sort of immigration policy. What policy is that. And why?

22. Which is more conservative, traditional, and dull, Australia or New Zealand?

AUSTRALIA AND THE SOUTH PACIFIC

1. In the 1960s and 1970s, there was a rock band called Three Dog Night—remember "Jeremiah Was a Bullfrog" and so forth. The rockers must have known of the aborigines' description of nights so cold a person had to huddle next to three dogs to keep warm. That ain't no way to have fun, as lyrics to one of the group's songs say.

2. New Brunswick is not an Australian state.

3. The Australian national anthem is "Waltzing Matilda."

4. Ninety percent of Australians live in or near the big coastal cities of Sydney, Brisbane, Melbourne, Adelaide, and Perth. Almost nobody lives in the famous "outback," although 130 million sheep do—about nine for each Australian.

5. **a** Kiribati used to be known as the Gilbert Islands. There are sixteen major islands here, with a total land area of only 291 square miles and a population under 60,000. But they have 800 square miles of lagoons, and these provide a fine livelihood for the people, who are Polynesians. Veterans of World War II may remember the name of one Kiribati island: Tarawa.

   b Vanuatu was called the New Hebrides, for which the French and English once shared responsibility. Its largest island is Espíritu Santo, but the capital, Vila, is on Efate. Vanuatu has 5,714 square miles of land and about 120,000 people, chiefly Melanesians, but there are some Europeans and Vietnamese.

   c Tuvalu is the former Ellice Islands, in eastern Polynesia, ten low coral atolls with a total area of ten square miles and 7,500 people. The airport baggage ticket of Funafuti island is abbreviated "FUN."

6. Mostly Pacific Ocean, at least for a long, long way. Nearly all of North Island is east of South Island, not north of it, so that South Island is not south of North Island . . . Oh, never mind: Look at the map.

7. New Zealand has about 68 million sheep and 3.2 million people, or about twenty-one to one, a statistic that illuminates the importance of sheep and sheep products to the country's economy.

8. Melbourne is the third largest Greek city.

9. Victoria and New Jersey are both officially known as the Garden State.

10. Swift's creatures were Yahoos, a term that has entered the English language, meaning something like "vulgar louts."

11. **a** *To barrack for* is "to cheer for your team"
   b A *bluey* is "a person with red hair"
   c *Bonzer* means good
   d *On the outer* means "broke"
   e *Tucker* is "food"
   f A *digger* is "an Australian soldier"

12. Twenty percent of all Australians live in or near the four shires, thirty-three municipalities, and seven cities that comprise Sydney, New South Wales, founded in 1788.

13. The original Australians, the aborigines, scarcely participate in modern Australian society. Of the 350,000 or so who lived here when the Europeans arrived, only 50,000 survive. A few work as janitors in the cities, more live on the dole near missions, and many of the rest work on sheep and cattle stations, as ranches are called. In only a few places do small groups pursue their original way of life.

14. A sentence of "transportation" meant deportation, likely to a prison colony in what is now New South Wales, Australia. If you survived a boat trip as harrowing as the ones African slaves had been subjected to, after serving your sentence you might find yourself banished to a wonderful new land of clear blue skies and unlimited opportunity, quite unlike the filthy, noisy, crowded, crime-ridden, sewers-in-the-streets society that had transported you. About 160,000 such deportees had arrived in Australia by the middle of the nineteenth century, and they and their descendants have built the country.

15. Perth is about as far from Sydney as L.A. is from Atlanta.

16. Australia's present-day population of 15 million is about the same as that of the United States in 1825.

17. The federal capital of Australia is Canberra, a new city designed and built in 1927, about halfway between Melbourne and Sydney. It is the only major Australian city that is not on the coast.

18. What ended the need for subsidized immigration into Australia was the discovery of gold. In a single decade, the populations of some areas increased tenfold. Forty percent of the world's gold came from Australia in the 1850s, and prosperity reigned.

19. Lead and zinc are the main products at Broken Hill, which was discovered after the gold rush wound down.

20. Australia has 5.5 million passenger cars and 15 million people, one of the highest ratios in the world.

21. The so-called White Australia policy encouraged immigration only by whites. Asians and blacks would have competed on the job market. Recently, the needs of Asian boat people and world public opinion have pressured the Australians for change.

22. New Zealand is the duller, most would agree. It is losing population, especially the young, to the more brash, ambitious, and hell-raising Australia.

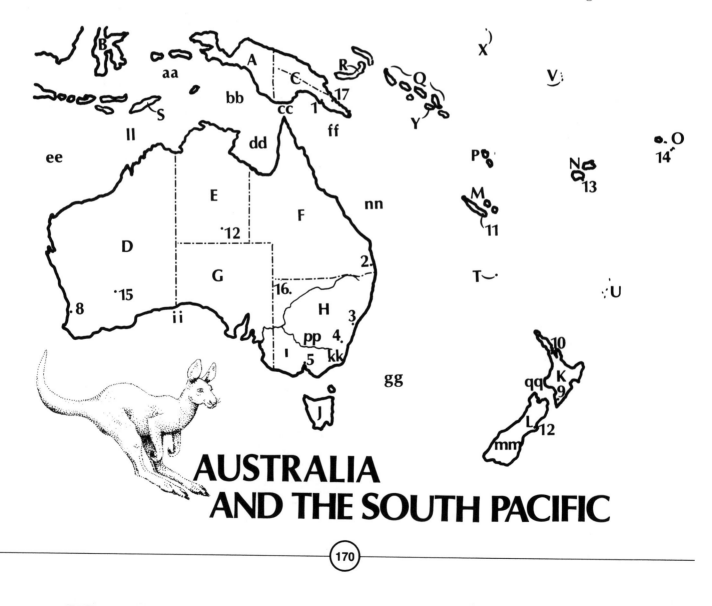

AUSTRALIA
AND THE SOUTH PACIFIC

AUSTRALIA AND THE SOUTH PACIFIC

Locate the following on the accompanying map of Australia and the South Pacific (countries, states, island chains, and important islands are marked with letters; cities with numbers; rivers, seas, straits, gulfs, and mountains with double letters):

Great Barrier Reef _____
Tasmania _____
South Island _____
Western Australia _____
Papua New Guinea _____
New Caledonia _____
Fiji _____
Queensland _____
New South Wales _____
Brisbane _____
Melbourne _____
Sydney _____
Canberra _____

Auckland _____
Wellington _____
Nouméa _____
Perth _____
Irian Barat _____
Gulf of Carpentaria _____
Tasman Sea _____
Pago Pago _____
Solomon Islands _____
Torres Strait _____
Timor _____
Great Australian Bight _____
Murray River _____

Vanuatu _____
Tuvalu _____
Coral Sea _____
Celebes _____
New Britain _____
Guadalcanal _____
Norfolk Island _____
Kermadec Islands _____
Nauru _____
Port Moresby _____
Broken Hill _____
Great Dividing Range _____
Southern Alps _____

AUSTRALIA AND THE SOUTH PACIFIC MAP QUIZ

Great Barrier Reef **nn**	Auckland **10**	Vanuatu **P**
Tasmania **J**	Wellington **9**	Tuvalu **V**
South Island **L**	Nouméa **11**	Coral Sea **ff**
Western Australia **D**	Perth **8**	Celebes **B**
Papua New Guinea **C**	Irian Barat **A**	New Britain **R**
New Caledonia **M**	Gulf of Carpentaria **dd**	Guadalcanal **Y**
Fiji **N**	Tasman Sea **gg**	Norfolk Island **T**
Queensland **F**	Pago Pago **14**	Kermadec Islands **U**
New South Wales **H**	Solomon Islands **Q**	Nauru **X**
Brisbane **2**	Torres Strait **cc**	Port Moresby **1**
Melbourne **5**	Timor **S**	Broken Hill **16**
Sydney **3**	Great Australian Bight **ii**	Great Dividing Range **kk**
Canberra **4**	Murray River **pp**	Southern Alps **mm**

ANTARCTICA

Antarctica is a strange place. It's the only spot on earth where the Russians allow unannounced visits from Americans inspecting their research stations; where there are wildlife preserves no airplane is allowed to fly over so as to protect the environment from pollution by exhaust fumes; where charged particles from the sun strike rarefied ionospheric gases and make fluttering auroras of light; where . . . let's see how much you know:

1. One statement in the preceding introduction is untrue. Which is it?

2. We state categorically that penguins are terrific. How they can live in such a place and look so great, and have such a wonderful sense of humor, we don't know. But they do, and we admire them for it. Nothing makes us laugh like a swimming penguin leaping straight up out of the water onto an ice floe. A swimming penguin can leap *(three, seven, nine)* feet straight out of the water. Choose one.

3. Of the Arctic and the Antarctic, which would you say has more ice? Why? (It has to do with the essential difference between them.)

4. There were thirty-eight scientific research bases in Antarctica in 1983. Which country had the most?
France
U.S.A.
USSR
Argentina
Japan
Poland

5. The Ross Ice Shelf is one of the great features of Antarctica. How big is it:
The size of Rhode Island
The size of the southeastern United States
The size of Australia
The size of Ireland

6. Is Antarctica completely covered by ice?

7. How many nations are active in Antarctica?
  a Five
  b Sixteen
  c Twenty-nine
  d Forty-four
  e Seventy-one

8. When U.N. representatives of the Third World began to clamor for control of Antarctica for "the common heritage of mankind"—by which is meant "for the Third World"—what did Antarctic Treaty member nations do to fend them off? Note: These member nations include the United States and the USSR.

9. What is Antarctica's largest land animal? Hint: The penguin isn't considered a land animal.

10. Is the South Pole warmer or colder than the North Pole?

11. On a small island, just offshore from the Antarctic Peninsula, is a Polish polar research station. What is its name?

12. The mapmakers of Columbus's day had a three-word name for Antarctica. Fill in the final evocative word: Terra Australis _____.

13. Who was the first to use the airplane to explore the Antarctic?

14. If you were looking for a lot of nicely preserved meteorites, why would you look in Antarctica?

15. What makes it possible for an animal to live in such a hostile environment as the Antarctic:
Size
Shape
Insulation

16. What is the southernmost mammal?

17. Clarence Birdseye got the idea for frozen foods from polar exploration, and of course it works. Members of a 1955 expedition ate some bread left behind by Shackleton's expedition fifty years earlier and found it, "a little dry, but otherwise it tasted fine." Apparently, two factors together keep food frozen in the Antarctic fresh. One is cold. The other is _____.

18. Scientists living in the Antarctic for months at a time sometimes turn a little weird. One may get "the big eye." Another may get "the long eye." Can you define these terms?

19. A man doing hard physical work in a temperate climate needs about 3,500 calories a day. About how many does the same man working in the Antarctic need?
4,000
4,500
5,000

20. As anyone knows who has lived in a cold climate, one of the hazards there is touching cold metal with moist, bare hands. What is one advised to do if that happens in the Antarctic?

21. Why do lost people walk in a circle?

1. It is not true that the Antarctic is the only place that has an aurora; those dramatic light shows appear at both poles. The Arctic's aurora borealis is a more familiar term to us in the Northern Hemisphere than the Antarctic's aurora australis.

2. The penguin can jump at least seven feet straight up from the water. There are eighteen to twenty species of penguin, only two of which—the Adélie and the emperor—inhabit the Antarctic. Fossils suggest that a five-foot penguin weighing 250 pounds once lived here.

3. The Antarctic has about eight times more ice than the Arctic. The former is a continent; land conserves heat poorly, so ice forms more easily and remains longer. The Arctic is an ocean, which can more easily store the summer's heat. Arctic ice is often thinner than thirty feet, but the sheet of ice covering most of the Antarctic is more than a mile thick.

4. Argentina, which considers Antarctica to be a sort of polar annex, not long ago had more (nine) research stations in Antarctica than any other country. The USSR was second, with seven; and the United Kingdom third, with five; followed by the United States (four); Chile (three); Australia (three); Japan (two); and France, South Africa, West Germany, Poland, and New Zealand, each with one. The truth is there may not be all that much science to be done in Antarctica, but it is the next resource frontier, and pretending to be scientific while staking a claim is useful geopolitics. Antarctica contains proven deposits of nearly every important metal, as well as coal and oil. Offshore fisheries have been important for a century.

5. The Ross Ice shelf, like the Filchner Ice Shelf, is a great raft of ice that projects into the surrounding sea and sits on the Antarctic continental shelf. It is about the size of the southeastern United States.

6. Antarctica is almost, but not quite, completely ice-covered. The peaks of the Transantarctic mountain range are not always covered—an interesting reversal of the idea of snow-capped mountain peaks. There are also ice-free areas near the Ross Ice Shelf and on the Antarctic Peninsula.

7. Sixteen countries are signatories to the Antarctic Treaty agreeing to the demilitarization of the continent south of 60 degrees latitude and have scientific bases or do ongoing research there.

8. The Antarctic Treaty member nations employed what Jeff Wheelwright calls "a novel tactic," which "was to send those U.N. delegates responsible for the clamor directly to Antarctica, so that they might see how forbiddingly impractical the development of resources would be." In a Quonset hut on Beardmore Glacier, the chief Soviet delegate stared at the visiting Malaysians and Tunisians, pounded on a tabletop, and called attacks on the treaty "objectively reactionary."

9. The largest land animal in this frozen place is a wingless insect less than a tenth of an inch long and related to the housefly. Yet in the icy waters surrounding Antarctica live many mammals, including the blue whale, the largest mammal.

10. At the South Pole, the temperature is often 60 degrees F below freezing, and nine thousand feet of ice cover the bedrock of the continent. A fifteen-mile-per-hour wind blows almost constantly across this flat, pale blue, featureless place. The North Pole is often warmer by 50 degrees, and more difficult to locate, because it lies in a jumble of shifting, grinding pack ice.

11. The Polish polar research station is called—what else?—Arctowski.

12. To the old mapmakers, Antarctica was known as Terra Australis *Incognita*, or "unknown."

13. Admiral Richard E. Byrd was first to explore this continent with the airplane. Amundsen had taken ninety-nine days to reach the pole and return in 1911; Byrd flew there from Little America, an island base within the Ross Ice Shelf, and back in a day.

14. Meteorites fall everywhere about equally, but those that fall in Antarctica fall on the glaciers, and nobody bothers them; as the glaciers ooze and slide forward, meteorites are carried to the glaciers' terminus, where they pile up as good as new.

15. Insulation is what makes survival possible, and this is why humans cannot live here without much artificial support. The most common form of insulation is two-layered: a layer of feathers or fur just over the skin and a layer of fat or oily tissue just beneath the skin. Together, they retain the body's heat.

16. The Weddell seal is the mammal living farthest south. Another seal, the ringed seal, is the mammal dwelling farthest north. Seals can stay submerged in icy water for as long as twenty minutes.

17. Cold and the absence of bacteria together preserve food in the Antarctic. There is only about one bacterium per pint of snow.

18. Someone has "the big eye" when he sits up sleepless for hours at a time. He has "the long eye" when he stares but doesn't see. As a sailor said, it's "the long eye" when "they have a twelve-foot stare in a ten-foot room."

19. A man working here needs about 5,000 calories a day.

20. The U.S. Navy's *Polar Manual*, written by a surgeon, Captain E. E. Hedblom, offers this advice: "Do not touch cold metal with moist, bare hands. If you should inadvertently stick a hand to cold metal, urinate on the metal to warm it and save some inches of skin. If you stick both hands, you'd better have a friend along."

21. Captain Hedblom thinks lost people walk in a circle because we are all asymmetrical, and a "solitary man probably circles in the direction of his shorter leg."

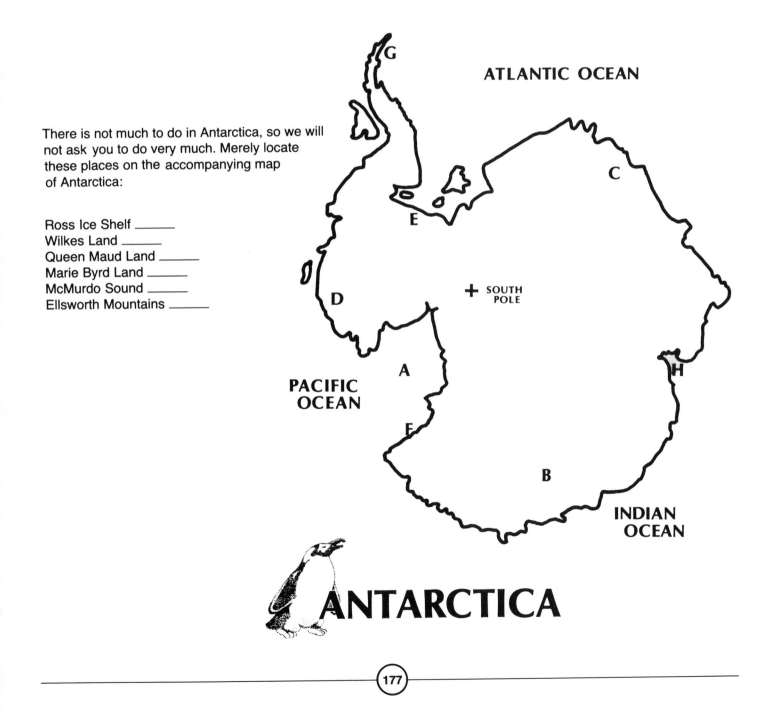

ANTARCTICA

There is not much to do in Antarctica, so we will not ask you to do very much. Merely locate these places on the accompanying map of Antarctica:

Ross Ice Shelf _____
Wilkes Land _____
Queen Maud Land _____
Marie Byrd Land _____
McMurdo Sound _____
Ellsworth Mountains _____

ATLANTIC OCEAN

G

C

E

D

+ SOUTH POLE

H

PACIFIC OCEAN

A

F

B

INDIAN OCEAN

ANTARCTICA

ANTARCTICA MAP QUIZ

Ross Ice Shelf **A**
Wilkes Land **B**
Queen Maud Land **C**
Marie Byrd Land **D**
McMurdo Sound **F**
Ellsworth Mountains **E**

IV

NORTH AMERICA

CANADA

O Canada, land of the hockey puck, land of lots of dark green trees up there in the glacial wastes, and that's about it, right? O how the Canadians detest that view we have of them, as a kind of pale, underdeveloped homuncular attachment to us, as bumpkins somehow foreign but not enough different from ourselves to be exotic or even interesting, as a people drawn away from England and snapped off, like ourselves and the Australians, but not powerful and smug like ourselves and not crazed with hairy-chested outback beer-drinking life-lust, like the Australians. Just kind of quiet, boring neighbors to the north, too nice to make enough trouble to be noticed. If Canadians were one of the less savory Third World peoples, this kind of insult could turn them to terrorism, it infuriates them so (in their pale, courteous way).

Yet, they go on being civil to us, breathing our acid rain, beating us at hockey, passively watching our dumb television programs (most Canadians live within one hundred miles of the U.S. border, and, Lord help them, we come in loud and clear).

The fact is, Canada is a great country—the world's second largest—and is only beginning to make its place in world history. And geographically, Canada is more worldly than we think. Montreal is located closer to Paris than Dawson City, and Halifax is nearer to Caracas than to Calgary. Canada has intelligent people, astounding amounts of resources, and civility.

1. What familiar place name in Canada means "shining waters"?

2. Joni Mitchell, Gordon Lightfoot, Alexander Graham Bell, and Hiram Walker are all famous Canadians—true or false?

3. Which Canadian province is officially bilingual?

4. Match these Canadian national parks with their provinces or territories:

1 Auyuittuq
2 Banff
3 Cape Breton Highlands
4 Forillon
5 Georgian Bay Islands
6 Gros Morne
7 Kluane
8 Prince Albert
9 Yoho

a Alberta
b Nova Scotia
c British Columbia
d Saskatchewan
e Yukon
f Newfoundland
g Ontario
h Quebec
i Northwest Territories

5. What does *Canada* mean?

6. The southernmost point in all Canada is in southwestern Ontario. How many U.S. states have land further north than that?

7. Where is Cameron Island, and what happened there recently that you should probably care about?

8. We tend to think of the United States as having two or three regions—either east and west, or east, middle, and west. Think about it, and you'll see why that's a geographic distinction based on climate, topography, and history. How many such regions is Canada most often divided into, and what are they?

9. What river did the French use to explore and settle Canada?

10. Who "owned" most of Canada in 1713?

11. What is the Palliser Triangle?

12. This part of southern Quebec, called the _____ _____, produces many good hockey players.

13. What and where is "the golden horseshoe"?

14. There is a thirty-foot Canadian five-cent piece standing outside Sudbury, the world's largest nickel mine. What is this giant nickel made of?

15. Name Canada's ten provinces and two territories.

16. What does "swastika" have to do with Canada?

17. A well-known epic poem describes the forced evacuation of Acadians in 1755. What is it?

18. What is Canada's newest province?

19. Why is Maine called "Down East"?

20. Many places have famous "walled cities"—the Forbidden City in China, Rothenburg in Germany, to mention two very different kinds. What is the only "walled city" in North America?

21. What strait divides Vancouver Island from the rest of British Columbia?

22. Which is the Sandstone City?

23. If you remember what *mestizos* are from an earlier quiz, you will be able to figure out, if you don't already know, what *metis* are. What are they?

24. One American president had his summer "White House" outside the United States. Who was he, and where was it?

CANADA

1. The word *Ontario* means "shining waters," and that province has nearly 69,000 square miles of freshwater surface—about a quarter of the earth's total.
2. True, all are Canadian.
3. The official language of Quebec is French. New Brunswick is the bilingual province of Canada, where the provincial government does business in both French and English. Canada's national government is also officially bilingual.
4. 1–i 6–f
   2–a 7–e
   3–b 8–d
   4–h 9–c
   5–g
5. The origin of the word *Canada* is not known for certain, but there are two strong candidates. It may derive from the Huron-Iroquois word *kanata*, meaning "village." Or, it may have been the exclamation of some Portuguese sailors, who arrived with the Gaspar Corte-Real expedition in 1500. "*Ça nada*," uttered with a disgusted inflection, means "Here, nothing!"
6. Surprisingly, no fewer than twenty-seven U.S. states have land north of Canada's southernmost point.
7. Cameron Island is in the Canadian Arctic, in the Parry Islands. Oil has recently been discovered there. Advanced new techniques of extraction are being tested there. If they succeed, then the oil of the Antarctic could also be extracted and delivered, setting off the first high-tech resource race of the 1990s.
8. Canada divides itself into two parts: the north, where it is cold and sparsely populated; and the south, where it is cold, but less so, and also sparsely populated, but less (sparsely) so.
9. The French used the Saint Lawrence for exploring Canada.
10. The Hudson's Bay Company, whose traplines and claims were more than twice as large as those of the French.
11. The Palliser Triangle is a large section of southeastern Alberta and Saskatchewan that is relatively dry and was avoided when the prairie provinces were settled. The name comes from that of Capt. John Palliser, who explored and surveyed the prairie provinces for the British government in 1857–60.
12. Eastern Townships.
13. The "golden horseshoe" is a string of cities that line the western shores of Lake Ontario, extending from northeast of Toronto to southeast of Hamilton.
14. Sudbury's nickel is made of stainless steel, which makes sense in Sudbury, a place notorious for its acidic polluted air and mountains of mine and refinery slag. This stainless-steel nickel will go a long way.
15. The territories are Northwest Territories and Yukon. The provinces are Prince Edward Island, Newfoundland, New Brunswick, Nova Scotia, Quebec, Ontario, Manitoba, Saskatchewan, Alberta, and British Columbia.
16. Swastika is the name of a small town in northern Ontario.

17. *Evangeline*, by Henry Wadsworth Longfellow, is about the expulsion of the Acadians, who went to Louisiana in the United States and became Cajuns. The poem and the poet are much thought of by Acadians and by Cajuns.

18. Canada's newest province is Newfoundland, which is also its oldest. First permanently settled in 1583, Newfoundland was an independent dominion from 1855 to 1934. Then bankruptcy forced a return to colonial status, and Newfoundland joined Canada—after a bitter referendum—on March 31, 1949.

19. You can see why Maine is called Down East if you look at it from Montreal and other parts of Quebec, where the term originated. To Quebec, Portland's year-round port has become a vital link for many imports and exports.

20. The only walled city in North America is Quebec City. Walls have stood around it since 1620, when Champlain built his fort there. The fortifications have been refurbished extensively in this century.

21. The Strait of Georgia divides Vancouver from the rest of British Columbia. *Georgia Straight* was the name of Vancouver's alternative newspaper during the early 1970s.

22. After Calgary, Alberta, burned down in 1886, only eleven years after it was founded, sandstone was mandated for subsequent building construction. Calgary's downtown nowadays is made of the usual steel-and-glass towers, but many of the old sandstone buildings remain.

23. *Metis*—people of French-Indian ancestry—are the *mestizos* of the north. Many of Canada's early trappers, lumbermen, and explorers of the interior were *metis*.

24. FDR's family spent summers on Campobello Island in New Brunswick.

CANADA

CANADA

Locate the following on the accompanying map of Canada (provinces are marked with capital letters, cities with numbers, rivers and mountains with double letters):

Quebec (Province) _____
Ontario _____
New Brunswick _____
Nova Scotia _____
Newfoundland _____
Prince Edward Island _____
Saskatchewan _____
Manitoba _____
Alberta _____
British Columbia _____
Yukon _____
District of Mackenzie _____

Quebec City _____
Toronto _____
Montreal _____
Halifax _____
Windsor _____
Winnipeg _____
Regina _____
Calgary _____
Edmonton _____
Vancouver _____
Victoria _____
Whitehorse _____

Fraser River _____
Peace River _____
Vancouver Island _____
James Bay _____
Gulf of Saint Lawrence _____
Victoria Island _____
Lake Superior _____
Lake Erie _____
Lake Huron _____
Queen Charlotte Islands _____
Baffin Island _____
Mackenzie River _____

CANADA MAP QUIZ

Quebec **A**	Quebec City **1**	Fraser River **cc**
Ontario **E**	Toronto **6**	Peace River **ff**
New Brunswick **D**	Montreal **2**	Vancouver Island **aa**
Nova Scotia **C**	Halifax **3**	James Bay **pp**
Newfoundland **B**	Windsor **7**	Gulf of Saint Lawrence **ll**
Prince Edward Island **F**	Winnipeg **13**	Victoria Island **nn**
Saskatchewan **H**	Regina **10**	Lake Superior **gg**
Manitoba **G**	Calgary **11**	Lake Erie **jj**
Alberta **I**	Edmonton **12**	Lake Huron **ii**
British Columbia **J**	Vancouver **14**	Queen Charlotte Islands **bb**
Yukon **K**	Victoria **15**	Baffin Island **mm**
District of Mackenzie **L**	Whitehorse **17**	Mackenzie River **dd**

THE U.S.A.

The U.S.A. is that famous nation of North America whose people love geography when they get a chance to learn some of it, but do not often get the chance. This is puzzling, since geography is not only fascinating, it is useful, and we are a practical people.

But even a practical people cannot always see the importance of knowing where the Sahel is, or what country produces the most tin or cacao, or where the panda lives and what it eats, or what part of the world still has many active volcanoes. We don't always bother to try to see things whole.

Perhaps we should, though, rather than wait until the Sahel desert begins to spread at a mile a day and hundreds of thousands are starving, or until we run short of tin or oil or chocolate (God forbid!), or until the panda is endangered, or until (surprise!) a Mount St. Helens erupts right here in our own country.

This is probably just human, not to bother with remote things until we are shown their immediate relevance.

But at least we know our own country, don't we?

1. Where does the sun first strike the U.S.?
Guam
Mount Desert Island
Mount Katahdin
Mount Tamalpais

2. What river did the Spanish use to explore and settle the United States?

3. In 1860, the greatest concentrations of slaves in the United States were in just four areas. Where were these?

4. Which is larger—the province of Quebec or the state of Alaska?

5. Both the United States and Canada are what is called "pluralistic societies." But their style of pluralism differs. Can you describe the difference?

6. As you move southward on the Atlantic coast, the landscape changes around Staten Island, New York. How does it change, and why the difference?

7. A hundred years ago, more than 70 percent of Americans lived in the country, outside cities and towns, most of them cultivating the land or raising stock. About what percentage of Americans are farmers today?
3 percent   29 percent
16 percent   49 percent

8. a Over the span of U.S. history, which country has provided the most immigrants?
b In the United States, the descendants of immigrants from what foreign country have attained the highest income and educational levels?

9. What do Truman Capote, Raquel Welch, and Martin Sheen have in common?

10. What do Larry Csonka, Joe Namath, and Don Shula have in common?

11. What do Redd Foxx, Moms Mabley, and Anita Bryant have in common?

12. In which American city would you find the most cars per capita?

New York   Miami
Chicago   Casper, Wyoming
Los Angeles

13. About how many American adults are illiterate?

2 percent   30 percent
14 percent   55 percent

14. Which five states: had the highest birthrates in 1980, and who lives there?

15. About how old is the average American?

Nineteen   Fifty-one
Thirty-one   Fifty-nine
Thirty-nine

16. True or false: Most Americans over eighty-five are in nursing homes.

17. Which state leads the country in executions since 1978?

18. Name the top two states in abortions in 1980?

19. How are ordinary U.S. federal highways numbered?

20. How are U.S. interstate highways numbered?

21. A staple for the Okies on their way west to California during the Dust Bowl days was "Hoover hog." What was it actually?

22. Where in the United States is the most acidic acid rain?

23. Where does the United States assemble most of its thermonuclear bombs?

24. Which president established the first U.S. National Wildlife Refuge? Where was it?

25. Where is Gay Head?

26. Where is the German Coast?

27. Where is the Natchez Trace?

28. Why is Dixieland called Dixieland?

29. Match these television series with the locales in which they were set:

1	*Dragnet*	a	Cincinnati
2	*Ironside*	b	North Carolina
3	*Mike Hammer*	c	Florida
4	*Mayberry RFD*	d	Los Angeles
5	*The Untouchables*	e	New York
6	*Flipper*	f	Minneapolis
7	*Cheers*	g	San Francisco
8	*WKRP*	h	Chicago
9	*Mary Tyler Moore Show*	i	Boston

30. What is the most common county name in the United States?

31. Where is the National Hobo Convention held?

32. Who are the richest Indians in North America, and where do they live?

33. Here is a sort of connect-the-dots map of the United States. Can you tell what it represents?

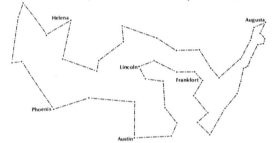

34. Where would you find North America's Danish Disneyland?

35. Can you identify the Iditarod, one of sport's longest distance athletic events?

36. What discovery in Alaska began immediately to change Alaskan life-styles and alter that state's environment forever?

37. About how many American men will serve time in state prisons?

a 1 percent   c 10 percent
b 5 percent   d 15 percent

38. Where is the Delmarva Peninsula?

39. What is the North American continent made of? Hint: This is the "magma" question (see introduction)!

THE U.S.A.

1. The sun first strikes the United States on Mount Katahdin, Maine.

2. Spanish explorers reached the Mississippi's mouth by 1500 and used that river to explore the new continent.

3. The slaves of America were concentrated largely in (a) the Atlantic coastal counties of Georgia and South Carolina (b) the Virginia counties along the Chesapeake Bay; (c) west-central Alabama in the "black soil" counties; and (d) the Mississippi River Valley from Memphis to Baton Rouge. Many areas of the south had very few slaves.

4. We were surprised to learn, as perhaps you will be, that Quebec is larger than Alaska—594,000 square miles versus 586,000.

5. Pluralism in the United States is mainly ethnic—various ethnic groups live side by side in most cities and cross over in hundreds of ways in daily life. In Canada, there are essentially two different language groups—English-speaking and French-speaking—and they live in different regions.

6. North of New Jersey, the U.S. coast is rocky and noisy with pounding surf—think of the coast of Maine. From New Jersey south, the coast is sandy beach until you reach the Florida Keys, where the limestone of coral islands and the tropical climate changes everything again. Why the difference? The glaciers of the Ice Age drove their icy fingers as far south as the present Staten Island, more or less, and then stopped. They created the shape of New England, scouring the land and heaping up moraines, such as Long Island and Cape Cod.

7. Only about 3 percent of Americans are farmers today. But they produce plenty of food—with surplus enough to export many kinds abroad. The productivity of the American farm, in fact, is the envy of the world. It is achieved at the social cost of the death of the family farm, the growing economic power of farm conglomerates and mechanized farming, and in some cases, a drop in the aesthetics and flavor of the food.

8. a Germans came in the greatest numbers, providing 14.8 percent of immigrants; Italians were next with 11.1 percent.

    b Russians have been most successful in America, according to *The Ethnic Almanac*. Many of these were Jews, of course, but the reception committee on Ellis Island asked people where they came from, not their religion (one reason people came here).

9. Capote, Welch, and Sheen are Americans of Spanish ancestry.

10. Csonka, Namath, and Shula are Americans of Hungarian descent.

11. Foxx, Mabley, and Bryant are Americans of American Indian descent.

12. In Casper, Wyoming, there are 729 cars for every 1,000 people.

13. According to Jonathan Kozol, a full third of American adults (60 million) are either completely illiterate or illiterate for all practical purposes in a technical society. Another source, a report by the Library of Congress based on a Yankelovich study, is less worrisome, claiming that a mere 23 million adult Americans are functionally illiterate. The number, whatever it is, is thought to be growing by more than two million per year. More than one-half—53 percent—are women, and most of them are poor. Of those adults who can read, 44 percent never read a book. Kozol, in his book *Illiterate America*, says that we should be furious about this, and frightened by it. We should, he says, attack the problem with an investment of $10 billion. He also says that we will not do so, and the results will someday be catastrophic. Only 2 to 4 percent of the illiterate adults are enrolled in literacy programs.

14. Utah (30.1 births per 1,000 people), Alaska (22.4), Idaho (22.1), New Mexico (20.6), Wyoming (21.7) are the five states with the highest birthrates. These are all rural states, with a large proportion of Mormons living in them, as well as many Catholic Mexicans. Mormons believe that having children is a sacred duty.

15. The average American is now 31.2 years old, according to Census Bureau figures, and getting older every year. The figure is up from 27.9 in 1970. And now, for the first time in history, there are more Americans over sixty-five than teenagers.

16. False. More than half of Americans over eighty-five are independent householders, almost 30 percent living alone. Fewer than a quarter live in nursing homes. Which is a good thing. Two out of three single elderly Americans who enter nursing homes are reduced to poverty within thirteen weeks.

17. Florida owns the dubious distinction of leading the United States in the number of executions since 1978, with 12 dead as of May 1985. It also leads the nation in the number of prisoners on Death Row—228— awaiting their meeting with Old Sparky, as the electric chair in Raiford State Prison is jovially called. Old Sparky, of sturdy oak, was built by prisoners.

18. California and New York led the United States in abortions in 1980. Washington, D.C. was close behind.

19. East-to-west federal highways are even-numbered, increasing from north to south. North-to-south highways are odd, increasing from east to west.

20. East-west interstate highways have even numbers that get higher as you move from south to north; north-south highways have odd numbers that get higher as you move from west to east. Those with triple digits are city bypasses and spurs. States number their highways any way they like.

21. Hoover hog was rabbit, jack or otherwise, eaten by hungry Okies, who blamed President Herbert Hoover for the Great Depression.

22. The worst of the acid rain extends across Pennsylvania, New York, and New Hampshire, just downwind from the major industrial cities of the Northeast. Acid rain—created when the nitrous oxides and sulphur dioxide in factory pollutants and auto exhaust are converted in the atmosphere to nitric and sulphuric acid, which then rains out again—has become serious only since 1970, killing forests and ending most life in some lakes. Now, most of eastern North America is seriously affected.

23. We put our nuclear bombs together in Amarillo, Texas. Parts come from all over the country.

24. Teddy Roosevelt, the outdoorsman president, established Pelican Island National Wildlife Refuge, off the coast of Florida, in 1903. Today there are hundreds of such refuges, protecting every habitat type.

25. Gay Head is at the west end of Martha's Vineyard.

26. The German Coast is along the west bank of the Mississippi, about thirty miles north of New Orleans. Alsatians settled here in 1721.

27. The Natchez Trace was an early American road, connecting Natchez, Mississippi, with Nashville, Tennessee, five hundred miles away. It served traders with an overland route back to Nashville after a downstream trip by raft, barge, or steamboat on the Tennessee, Ohio, and Mississippi rivers.

28. Before the Civil War, the Citizens Bank of Louisiana issued bilingual currency. Because of inflation, so many of its ten-dollar bills were circulating (*dix* means "ten" in French) that the South came to be called Dixie.

29. 1–d   6–c
    2–g   7–i
    3–e   8–a
    4–b   9–f
    5–h

30. Thirty states have a Washington County, and Louisiana has a Washington Parish.

31. Britt, Iowa, every August when it's warm outside.

The hoboes—not tramps or bums or street-people or homeless wanderers, but hoboes—elect a king and queen of the road.

32. The Aguas Calientes of Palm Springs, who number less than two hundred individuals, are probably the richest Indian tribe. Their lands, held in a tribal trust, may be worth $1.5 billion.

33. The connect-the-dots map of the United States represents what *Discover* magazine called "a bravado exercise in combinatorial mathematics, involving some prodigious calculations." The result: By following this map, a salesman could visit forty-eight state capitals clocking the lowest possible mileage.

34. In Solvang, California, which is filled with Danish restaurants, stores, and simulated Danish buildings.

35. Iditarod is the dogsled race from Anchorage to Nome, held each year in March. The course varies from year to year. In 1982, it was 1,131 miles, and the top twenty finishers divided a purse of $100,000.

36. The discovery of oil, in early 1968, on Alaska's North Slope, turned out to be the largest oil field found so far in North America. Almost immediately, it began to change that state's pioneer image—and its reality. Trappers became oilmen, and oilmen came in the thousands from Oklahoma, Texas, and Pennsylvania. The Trans-Alaska Pipeline—the most expensive privately financed construction project in history, it was called—was the real beginning. Not everyone thinks this is great. The fragile Arctic could be damaged, environmentalists fear. And the isolation of the state's interior—one of the world's last frontiers—may be ended forever.

37. As many as 5.1 percent of adult American males is likely to do time, says the Justice Department. That is fourteen times the number of women expected to be imprisoned during their lifetimes in America. Between 11 and 18 percent of black men can be expected to do time, the Justice Department report says, and 2 to 3.3 percent of whites.

38. The Delmarva Peninsula is that enormous one enclosing the eastern side of Chesapeake Bay; it forms part of Virginia, Delaware, and Maryland. The inland portions of this region are well known for truck farming, and the shores for crabbing, clamming, and oystering. In this, as in much else, location is everything. The soils and climate here—light and mild, respectively—are perfect for vegetable farming. The bay, before it became polluted, seemed an endless source of seafood. And not far away are the huge markets of Baltimore, Washington, D.C., Philadelphia, and New York.

39. North America is bigger than it looks, because much of it is under the sea, spreading out from what we see on maps as much as one hundred miles on the Atlantic side. The igneous rocks beneath it all are granitics. Granite is a kind of rock. It is formed when another sort of rock—the hot, liquid rock found below the earth's surface and called magma (ahh!)—cools down and solidifies. All sorts of sediment overlay this from place to place, but the continent's core is granite, and the granite was once magma.

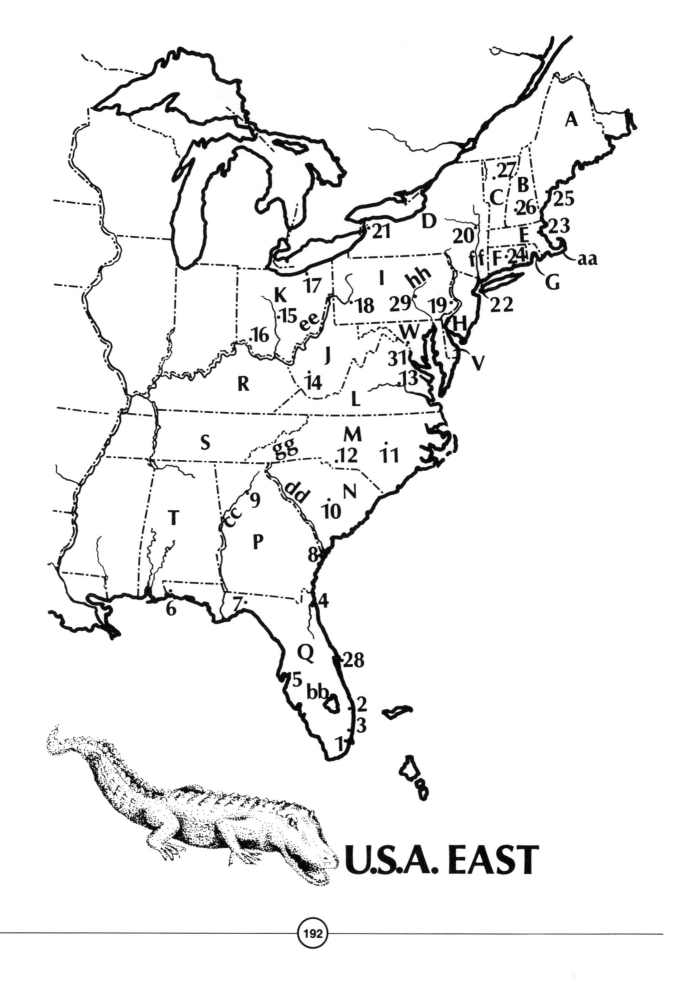

A

.27

B

C

.26

25

D

.21

20

E

23

ff F.24

aa

17

I

hh

G

K

.18

29

19.

22

.15

ee

W

H

.16

J

31

V

R

.14

.13

L

S

M

gg

.12

i1

dd

N

9

10

cc

T

P

8

6

7.

4

Q

.28

5 bb

2

3

1

U.S.A. EAST

U.S.A. EAST

Locate the following on the accompanying U.S.A. East Map (states are marked with letters; cities with numbers; rivers, lakes, mountains, and bays with double letters):

New York _____
Florida _____
North Carolina _____
Maine _____
Boston _____
Pennsylvania _____
Buffalo _____
New Jersey _____
Georgia _____
Cleveland _____
Ohio _____
Richmond _____
Connecticut _____

Vermont _____
Delaware _____
West Virginia _____
Cape Canaveral _____
Maryland _____
Miami _____
Harrisburg _____
Savannah _____
Rhode Island _____
Columbia _____
Cape Cod _____
Washington, D.C. _____
Albany _____

South Carolina _____
Palm Beach _____
Atlanta _____
New York City _____
Tallahassee _____
Concord _____
Great Smoky Mountains _____
Susquehanna River _____
Hudson River _____
Chattahoochee River _____
Charlotte _____
Jacksonville _____
Lake Okeechobee _____

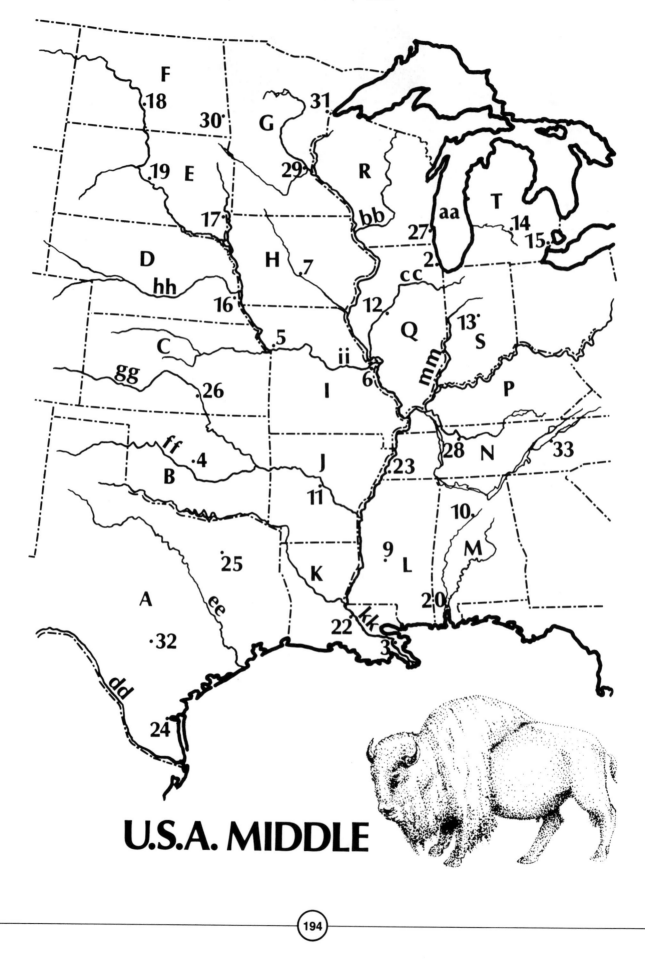

U.S.A. MIDDLE

U.S.A. MIDDLE

Locate the following on the accompanying U.S.A. Middle map (states are marked with letters; cities with numbers; rivers, lakes, and so on with double letters):

Texas _____
Illinois _____
Chicago _____
Minnesota _____
North Dakota _____
Lake Michigan _____
Tennessee _____
Alabama _____
Rio Grande _____
Mobile _____
Baton Rouge _____
Memphis _____
Corpus Christi _____

Oklahoma City _____
Dallas _____
Saint Louis _____
Des Moines _____
Omaha _____
Detroit _____
Bismarck _____
Kansas City _____
Wichita _____
Missouri River _____
Milwaukee _____
Illinois River _____
Indianapolis _____

Nashville _____
Kansas _____
Pierre _____
Wabash River _____
Arkansas River _____
Knoxville _____
Sioux Falls _____
Fargo _____
Duluth _____
Lansing _____
San Antonio _____
Brazos River _____

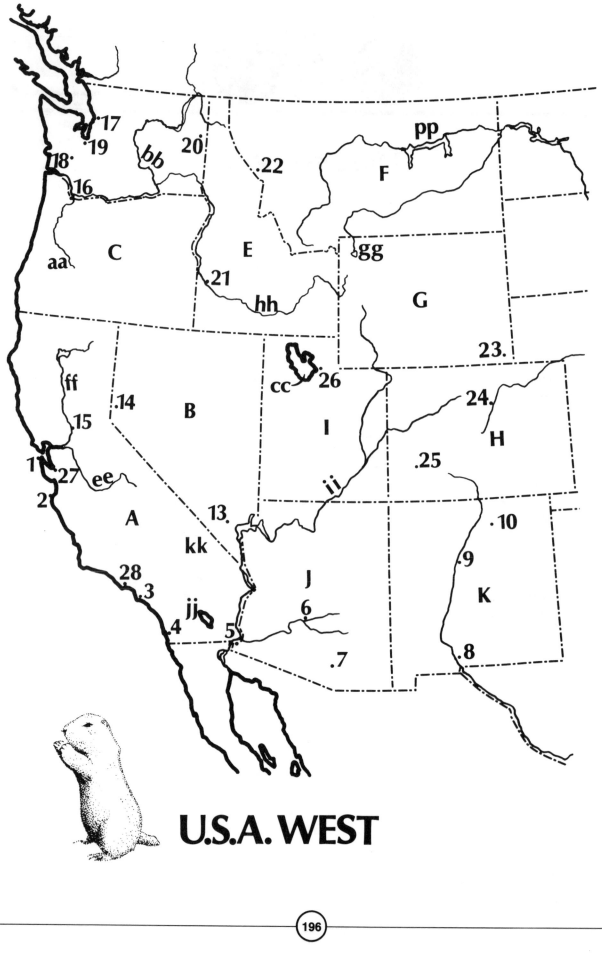

U.S.A. WEST

Locate the following on the accompanying U.S.A. West map (states are marked with letters; cities with numbers; rivers, lakes, and scenic locales with double letters):

California _____

Arizona _____

Colorado _____

Wyoming _____

Denver _____

Los Angeles _____

San Francisco _____

Portland _____

Idaho _____

Utah _____

Reno _____

Tucson _____

Colorado River _____

Phoenix _____

Snake River _____

Yellowstone National Park _____

San Diego _____

Albuquerque _____

Sacramento _____

Cheyenne _____

New Mexico _____

San Joaquin River _____

Montana _____

Spokane _____

Las Vegas _____

Boise _____

Death Valley _____

Malibu _____

Monterey _____

Willamette River _____

Olympia _____

Fort Peck Reservoir _____

Yuma _____

San Jose _____

Salton Sea _____

U.S.A. EAST MAP QUIZ

New York **D**
Florida **Q**
North Carolina **M**
Maine **A**
Boston **23**
Pennsylvania **I**
Buffalo **21**
New Jersey **H**
Georgia **P**
Cleveland **17**
Ohio **K**
Richmond **13**
Connecticut **F**

Vermont **C**
Delaware **V**
West Virginia **J**
Cape Canaveral **28**
Maryland **W**
Miami **1**
Harrisburg **29**
Savannah **8**
Rhode Island **G**
Columbia **10**
Cape Cod **aa**
Washington, D.C. **31**
Albany **20**

South Carolina **N**
Palm Beach **2**
Atlanta **9**
New York City **22**
Tallahassee **7**
Concord **26**
Great Smoky Mountains **gg**
Susquehanna River **hh**
Hudson River **ff**
Chattahoochee River **cc**
Charlotte **12**
Jacksonville **4**
Lake Okeechobee **bb**

U.S.A. MIDDLE MAP QUIZ

Texas **A**
Illinois **Q**
Chicago **2**
Minnesota **G**
North Dakota **F**
Lake Michigan **aa**
Tennessee **N**
Alabama **M**
Rio Grande **dd**
Mobile **20**
Baton Rouge **22**
Memphis **23**
Corpus Christi **24**

Oklahoma City **4**
Dallas **25**
Saint Louis **6**
Des Moines **7**
Omaha **16**
Detroit **15**
Bismarck **18**
Kansas City **5**
Wichita **26**
Missouri River **ii**
Milwaukee **27**
Illinois River **cc**
Indianapolis **13**

Nashville **28**
Kansas **C**
Pierre **19**
Wabash River **mm**
Arkansas River **gg**
Knoxville **33**
Sioux Falls **17**
Fargo **30**
Duluth **31**
Lansing **14**
San Antonio **32**
Brazos River **ee**

U.S.A. WEST MAP QUIZ

California **A**
Arizona **J**
Colorado **H**
Wyoming **G**
Denver **24**
Los Angeles **3**
San Francisco **1**
Portland **16**
Idaho **E**
Utah **I**
Reno **14**
Tucson **7**

Colorado River **ii**
Phoenix **6**
Snake River **hh**
Yellowstone National Park **gg**
San Diego **4**
Albuquerque **9**
Sacramento **15**
Cheyenne **23**
New Mexico **K**
San Joaquin River **ee**
Montana **F**
Spokane **20**

Las Vegas **13**
Boise **21**
Death Valley **kk**
Malibu **28**
Monterey **2**
Willamette River **aa**
Olympia **19**
Fort Peck Reservoir **pp**
Yuma **5**
San Jose **27**
Salton Sea **jj**

U.S.A. STATES

You might imagine that the states would need no introduction, and you're right. Go on to question 1.

1. People will name their towns just about anything, won't they? Granted, sometimes they're unimaginative, as in calling the place Johnson City or Greenville, or naming it after some patriotic figure, like Jefferson. But when they try, they can produce some amazing names. Here, for instance, is a list of American towns—we guarantee they actually exist. Can you fill in their states? (If you can name over ten, you're doing well.)

Needles, _____

Opp, _____

Alakanuk, _____

Bagdad, _____

Pocahontas, _____

Kaunakakai, _____

Chugwater, _____

Winooski, _____

Center Ossipee, _____

Rifle, _____

Spooner, _____

Mystic, _____

Woonsocket, _____

Locate, _____

Temple Terrace, _____

Sandwich, _____

Pocomoke City, _____

Dagsboro, _____

Unadilla, _____

Kooskia, _____

Cabool, _____

Paducah, _____

Alamo, _____

Mars Hill, _____

Rio Grande, _____

Loogootee, _____

Horseheads, _____

Roy, _____

Normal, _____

Coon Rapids, _____

Gackle, _____

Truth or Consequences, _____

Chagrin Falls, _____

Intercourse, _____

Bowdle, _____

Tooele, _____

Onancock, _____

Drain, _____

Duckwater, _____

Muleshoe, _____

War, _____

Klickitat, _____

Hominy, _____

Traveler's Rest, _____

Sedan, _____

New Iberia, _____

Paw Paw, _____

Yazoo City, _____

Wahoo, _____

Swan Quarter, _____

Blue Earth, _____

2. Where could you stand in four states at once, if you had four feet?

3. Only one state has no straight-line boundaries—can you name it?

4. Three states have *only* straight-line boundaries—can you name them?

5. Is Rhode Island an island?

6. Name the states in which these presidentially named places are situated. Note that having a presidential name does not necessarily mean you were named for a president.
Reagan County _____
Carter Dome _____
Ford City _____
Nixon (a town) _____
Johnson City (a town) _____
Kennedy Channel _____

Mount Eisenhower _____
Trumann (a town) _____
Roosevelt (a town) _____

7. Which state has the most national wildlife refuges?

8. What was the original name for the state that became Utah?

9. Can you name the largest county in the United States, or the state in which it is found? Or, for that matter, can you name a state that this largest county is sixteen times larger than?

10. There is only one state that borders only one state. Can you name it?

11. Only one state is entirely bordered by rivers to the east and west. Name the state—and the rivers.

12. Identify the following states from their silhouettes:

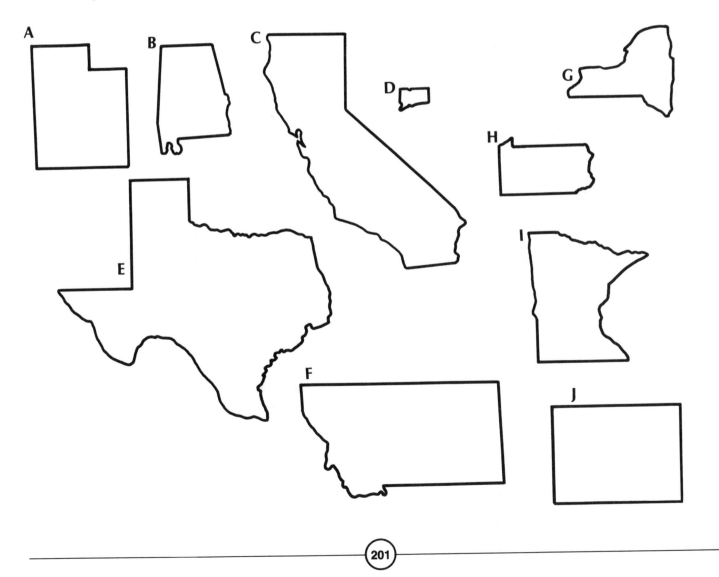

13. In which region, of which state, would you find a religious sect whose most conservative members refer to themselves as Old Order?

14. What state, excepting Alaska, has the greatest area of designated wilderness?

15. What state, excepting Alaska, has the most "roadless area"?

16. Why has one expert called Utah "darned near a Third World country in the middle of the United States"?

17. What state did Benjamin Franklin describe as "like a barrel tapped at both ends"?

18. What state was Sanford B. Dole president of? (That's right: president.)

19. Only five states have populations more than 25 percent black. Can you name them?

20. Only nine states have populations less than 1 percent black. Can you name them?

21. How long is a furlong, and why do we ask you that in a chapter on states?

22. Why did the Wright brothers, bicycle repairmen from Ohio, go to Kitty Hawk, North Carolina, to try out their airplane?

23. Texas is the third most populous state and still growing. Even so, it has more (_____?) than people.

24. What is the southernmost state in which you can ski most every winter?

25. Which state has the most people per square mile?

26. Which state has the fewest people per square mile?

27. What state has:
 a The most Civil War battlefields
 b The second most Civil War battlefields

28. Match these locations of Civil War battles with their states:
 1 Gettysburg   **a** Maryland
 2 Chickamauga   **b** South Carolina
 3 Bull Run   **c** Tennessee
 4 Antietam   **d** Virginia
 5 Shiloh   **e** Pennsylvania
 6 Fort Sumter   **f** Georgia

29. Five states lost population between 1980 and 1985. How many can you name?

30. Match these national parks with the states in which they are found:
 1 Bryce Canyon   **a** Arizona
 2 Everglades   **b** Colorado
 3 Yosemite   **c** Montana
 4 Crater Lake   **d** Washington
 5 Mount Rainier   **e** Oregon
 6 Glacier   **f** California
 7 Rocky Mountain   **g** Florida
 8 Grand Canyon   **h** Utah

31. Once upon a time, there was a state named Franklin, U.S.A. In what present state was that state?

32. Which state would you think has the most counties:
Texas
New York
Florida
Minnesota
Alaska
Hawaii
Louisiana
Wisconsin

33. People like them yellow, a recent study discovered, even though they don't have to be yellow for any particular reason—we just got used to yellow _____s. Fill in the blank, and tell what kind of wood is the best for making them, and where it comes from.

34. Name the modern states in which these famous Indian battles occurred:
Washita
Birch Coulee
Little Big Horn
Wounded Knee
Adobe Walls

35. What was the warmest temperature ever recorded in Alaska?
100 degrees F
87 degrees F
21 degrees F

36. What was the coldest temperature ever recorded in Florida?
35 degrees F
22 degrees F
Minus 2 degrees F

37. The United States may have more coal than anyplace else in the world. More than three-fourths of the country's coal production comes from three states. Can you name them?

38. What eastern state is known for its potatoes, blueberries, and lobsters?

39. Into what eastern state did many French-Canadian workers pour during the mid-nineteenth century to work in shoe and textile mills?

40. A midwestern state takes its nickname from the Cornish miners who dug into its hillsides and lived there during the winter. Can you give the state's name—and its nickname?

41. In which eastern state can be found the country's oldest newspaper in continuous publication?

U.S.A. States

1. The towns and their states are as follows:
Needles, California
Opp, Alabama
Alakanuk, Alaska
Bagdad, Arizona
Pocahontas, Arkansas and Iowa
Kaunakakai, Hawaii
Chugwater, Wyoming
Winooski, Vermont
Center Ossipee, New Hampshire
Rifle, Colorado
Spooner, Wisconsin
Mystic, Connecticut
Woonsocket, Rhode Island and South Dakota
Locate, Montana
Temple Terrace, Florida
Sandwich, Massachusetts and Illinois
Pocomoke City, Maryland
Dagsboro, Delaware
Unadilla, Georgia
Kooskia, Idaho
Cabool, Missouri
Paducah, Kentucky and Texas
Alamo, Tennessee, Georgia, and Texas
Mars Hill, Maine
Rio Grande, New Jersey and Puerto Rico
Loogootee, Indiana
Horseheads, New York
Roy, New Mexico and Utah
Normal, Illinois and Alabama
Coon Rapids, Iowa and Minnesota
Gackle, North Dakota

Truth or Consequences, New Mexico
Chagrin Falls, Ohio
Intercourse, Pennsylvania
Bowdle, South Dakota
Tooele, Utah
Onancock, Virginia
Drain, Oregon
Duckwater, Nevada
Muleshoe, Texas
War, West Virginia
Klickitat, Washington
Hominy, Oklahoma
Traveler's Rest, South Carolina
Sedan, Kansas
New Iberia, Louisiana
Paw Paw, Michigan
Yazoo City, Mississippi
Wahoo, Nebraska
Swan Quarter, North Carolina
Blue Earth, Minnesota

2. At a place called "the four corners," you could stand on Utah, Colorado, Arizona, and New Mexico all at once. It would take four feet to do it—or we suppose you could place half a foot on each state. We wouldn't consider that cheating.

3. The only state with no straight-line boundary is Hawaii. The shores of beaches are like that.

4. The three states with all straight-line boundaries are Wyoming, Colorado, and Utah.

5. Rhode Island is the smallest state, and it is also the name of the island on which Newport is located.
The state bird of Rhode Island is the chicken.

6. Reagan County is in Texas; Carter Dome is a mountain in New Hampshire; Ford City is in Pennsylvania (Ford City is also a shopping mall in Chicago); Nixon is in east-central Texas; Johnson City is in Blanco County, Texas; Kennedy Channel is along the northwest coast of Greenland, and not in the United States at all; Mount Eisenhower is in the Canadian Rockies; Trumann is in northeast Arkansas; and Roosevelt is in Alabama.

7. California has eight and is second. Florida, with nine, has the most national wildlife refuges.

8. Utah was first called Deseret and included most of the lands west of the Rockies and south of the forty-second parallel. At least this was what the Mormons proposed to Congress in 1849. Congress declined and created the Utah Territory the following year.

9. The largest U.S. county is San Bernardino County, east of Los Angeles. At 20,117 square miles, it is more than sixteen times bigger than Rhode Island. Actually, lots of counties are bigger than Rhode Island, which covers only 1,214 square miles, and San Bernardino County is bigger than lots of other states, including Delaware, Connecticut, Hawaii, New Jersey, Massachusetts, New Hampshire, Vermont, and Maryland.

10. Maine borders only New Hampshire.

11. Iowa is the only state with rivers to its eastern and western sides—the Mississippi and the Missouri—giving rise to such examples of wit as this remark from denizens of a neighboring state: "With good fences along the Missouri and Minnesota lines, they could run a lot of pigs over there in Ioway." For a landlocked state, Iowa has a lot of water. The Des Moines, Iowa, and Cedar rivers empty into the Mississippi, and the Sioux feeds the Missouri. French trappers and explorers were the first to visit Iowa, and they came by river.

12.
   a Utah
   b Alabama
   c California
   d Connecticut
   e Texas
   f Montana
   g New York
   h Pennsylvania
   i Minnesota
   j Wyoming

13. In the area of Lancaster, Pennsylvania, are the Mennonites, who take their name from a sixteenth-century Dutch Anabaptist leader, Menno Simons, and the Amish, named for Jacob Amman, who split from the Mennonites in 1693. The most fundamentalist members of both these sects call themselves Old Order.

14. Idaho has the most designated wilderness, with 3.86 million acres.

15. California has the most roadless area, with 19.4 million acres, and Idaho is second.

16. The expert in population studies compares Utah with a Third World country because the "baby boom" of the 1950s never ended in that state, where the birthrate is double that of the national average and approaches that of India. It shows no sign of slowing, either.

    Why? Because Mormons predominate in Utah.

17. Franklin was speaking of New Jersey and referring to the influence of New York at one end and Philadelphia at the other.

18. Dole was president of Hawaii in 1893, when it was a republic. Yes indeed, there was a connection between President Dole and the pineapples—he owned them.

19. Louisiana, Mississippi, Alabama, Georgia, and South Carolina have populations that are 25 percent black or more.

20. Idaho, Montana, Utah, Wyoming, North and South Dakota, Maine, New Hampshire, and Vermont have black populations under 1 percent.

21. A furlong is equal to ten chains, or forty rods, or an eighth of a mile. In the Kentucky bluegrass country, it is used to measure distances in horseraces.

22. The Wrights came to North Carolina because Kitty Hawk is on a barrier island near Cape Hatteras, where there is regular high wind to provide lift and no trees to run into.

23. Texas still has more *cows* than people.

24. Mount Lemmon, in the Santa Catalina Mountains north of Tucson, Arizona, is the southernmost ski resort in the United States.

25. New Jersey has the most, with an average of 1000. Rhode Island is second with 903.

26. Alaska has the fewest people per square mile, with less than one per square mile. Wyoming is next, with five.

27. a Virginia has the most Civil War battlefields.
    b Tennessee has the second most.

28.
   1–e 4–a
   2–f 5–c
   3–d 6–b

29. Iowa, Michigan, Ohio, Pennsylvania, and West Virginia lost population during this period. So did Washington, D.C. The decrease of 174,000 in Michigan was by far the largest. During 1980 to 1985, the population of the entire United States increased by more than twelve million.

30.
   1–h 5–d
   2–g 6–c
   3–f 7–b
   4–e 8–a

31. Franklin was established in 1784, in what later became eastern Tennessee. It was terminated in 1788. The capital was Greenville.

32. Texas has 254 counties, the most; and Georgia has 159, the second most.

33. You can't sell a pencil that isn't yellow, says Eberhard Faber Co., which has been making them since 1893 and now produces a million a day. When the company tried a varnished cedar pencil, which was beautiful and utilitarian, pencil dealers decided it was a mere fad and refused to buy it. Pencils are made of cedar, and the best cedar—best because it doesn't warp—comes from California. The lead isn't lead, as you probably knew; it's made of graphite mixed with clay.

34. Washita was in Oklahoma, Birch Coulee in Minnesota, Little Big Horn in Montana, Wounded Knee in South Dakota, Adobe Walls in Texas.

35. The warmest temperature ever recorded in Alaska was 100 degrees F, at Fort Yukon, on June 27, 1915.

36. The coldest temperature ever recorded in Florida was minus 2 degrees F, on February 13, 1899, in Tallahassee.

37. Pennsylvania, West Virginia, and eastern Kentucky have the nation's largest coal reserves. Another coalfield is found under western Kentucky and Illinois. There is also a lot of coal out west, particularly in Wyoming and Arizona, but it is of lower quality and far from markets, so it has only been tapped seriously in recent years.

38. Maine.

39. New Hampshire, a farming and seafaring state of English-speaking people until the mid-nineteenth century, needed more workers when the coming of railroads and clearing of timber created an industrial boom. French-Canadians answered the call, and their descendants are among the state's leading citizens.

40. Wisconsin—whose name derives from an Indian word, *misconsing*, meaning "grassy place," is sometimes called the Badger State.

41. In Connecticut is published the *Hartford Courant*. Also in Hartford, the nation's first cookbook was published, the pneumatic tire was devised, and once and for all, the standardized inch was established.

U.S.A. CITIES

Americans, more than any other people, seem to believe the essential virtues are honesty, integrity, work, thrift, and "naturalness," and that, moreover, these virtues can be found mostly in the country and their opposites mostly in the city. Yet most of us live in the city.

Why? Most would probably say, "To earn a living."

But there is more to cities than jobs. Cities are where one can disappear, cities are where there is excitement, variety, "night life," and human intelligence applied to the task of change, not merely to getting by. Cities— eminently human creations, human hives—are centers of crucial human activities, such as trade and innovation.

1. What are buffalo wings?

2. In 1790, what were the five largest U.S. cities?

3. Match the city with the festival in which it is held:

1 Mardi Gras	**a** Key West, Florida
2 Rose Festival	**b** Portland, Oregon
3 Frontier Days	**c** Pendleton, Oregon
4 Round Up	**d** Miami, Florida
5 Fantasy Fest	**e** Cheyenne, Wyoming
6 Orange Bowl	**f** New Orleans, Louisiana

4. A well-known Warsaw is in Poland. But Warsaws also exist in the United States, as do towns called:

Troy, _____
Utica, _____
Syracuse, _____
Versailles, _____

Paris, _____
Cairo, _____
Alexandria, _____
Memphis, _____
   Can you name the states they are found in?

5. Match the city with its nickname:

1 Big Apple	**a** Portland, Oregon
2 Queen City	**b** Boston
3 Crescent City	**c** Chicago
4 City of Brotherly Love	**d** New Orleans
5 City of Broad Shoulders	**e** Detroit
6 City of Angels	**f** Philadelphia
7 City of Roses	**g** Cincinnati
8 Motor City or Motown	**h** Los Angeles
9 Bean Town	**i** New York
10 Magic City	**j** Miami

6. Identify the cities depicted in each of the following street maps:

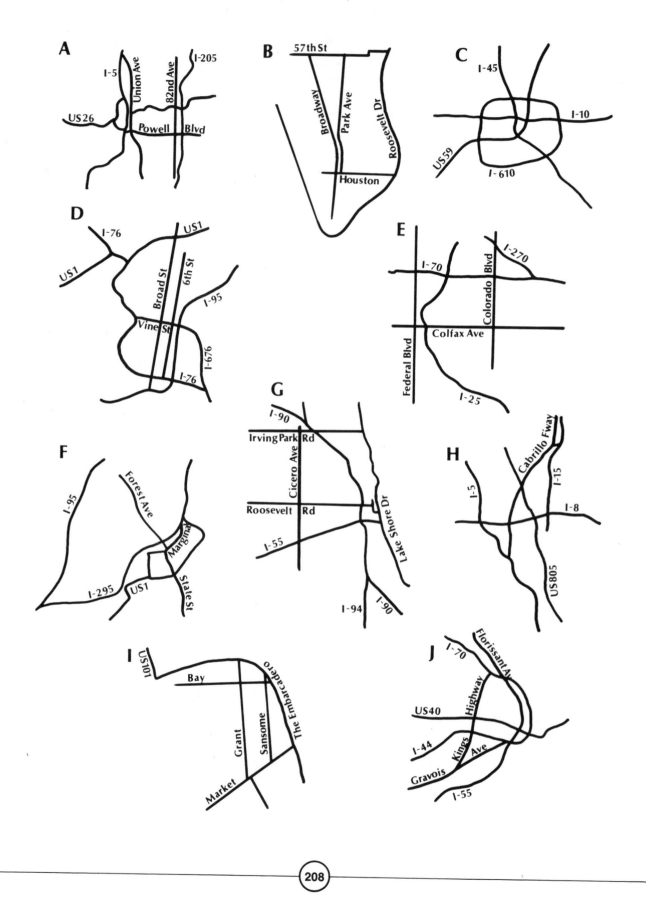

7. Of what city did Gertrude Stein say, against all the prevailing rules of geography, that when you get there, there's no there there?

8. Twenty thousand tons of *what* strike the New York City pavement every year?

9. What do these cities have in common: Virginia City, Nevada; Tombstone, Arizona; Leadville, Colorado; Coeur d'Alene, Idaho.

10. Where are Baltic Avenue, Mediterranean Avenue, and Marvin Gardens?

11. Which of these cities is on the Mississippi River?
  a Chicago
  b Memphis
  c Cincinnati
  d Kansas City
  e Des Moines

12. Which of these cities is on the Ohio River?
  a Cleveland
  b Detroit
  c Cincinnati
  d Columbus
  e Nashville

13. Which of these cities is on the Rio Grande?
  a San Antonio
  b Dallas
  c El Paso
  d Houston
  e Tucson

14. Which of these cities is on the Missouri River?
  a Des Moines
  b Omaha
  c Topeka
  d Minneapolis
  e Pittsburgh

15. Which of these cities is on Lake Michigan?
  a Milwaukee
  b Detroit
  c Cleveland
  d Madison
  e Indianapolis

16. Which of these cities is on Puget Sound?
  a Vancouver
  b Portland
  c Spokane
  d Walla Walla
  e Tacoma

17. Which of these cities is on San Francisco Bay?
  a San Jose
  b Sacramento
  c Fresno
  d Los Angeles
  e Stockton

18. Which of these cities is on the Colorado River?
  a Denver
  b Aspen
  c Salt Lake City
  d Las Vegas
  e Yuma

19. There are three state capitals on the Missouri River. Can you name two? Or three?

20. In what city or town did the man live who wrote, "The mass of men lead lives of quiet desperation"?

21. What town claims to be the birthplace of the Republican Party? Hint: It's in New York.

22. Match these educational institutions with their cities:
  1 Harvard   **a** Palo Alto
  2 Yale   **b** Hanover
  3 Brown   **c** South Bend
  4 Notre Dame   **d** Providence
  5 Dartmouth   **e** Cambridge
  6 Stanford   **f** New Haven

23. One U.S. city's telephone book has more than twenty-five pages of Johnsons and almost as many Andersons. Can you name the city?

24. Which U.S. city was established as "a holy experiment"?

25. What might "Blade 125" mean, in what city?

26. What was the original name of Moscow, Idaho?
  Minsk
  Henrietta
  Angel Town
  Saint Petersburg
  Hog Heaven
  Pinsk

27. There's no reason why you should know this, but take a guess what city the world's first coin laundry was in.

28. Match the sports arena with its city:

1	Maple Leaf Gardens	a	Pontiac
2	Wrigley Field	b	Houston
3	Shea Stadium	c	Inglewood
4	Orange Bowl		(suburban Los
5	Rose Bowl		Angeles)
6	County Stadium	d	Salt Lake City
7	Olympic Stadium	e	Seattle
8	Busch Stadium	f	Boston
9	Three Rivers	g	San Francisco
	Stadium	h	Minneapolis
10	Riverfront Stadium	i	Cincinnati
11	Silverdome	j	Pittsburgh
12	Hubert Humphrey	k	Saint Louis
	Metrodome	l	Montreal
13	Candlestick Park	m	Milwaukee
14	Fenway Park	n	Pasadena
15	King Dome	o	Miami
16	The Forum	p	New York City
17	The Salt Palace	q	Chicago
18	Astrodome	r	Toronto

29. What city is L'Enfant Plaza in, who is it named for, and why?

30. In 1893, George Ferris built something unusual in Chicago. What was it?

31. Match the city with the county in which it is situated:

1	Miami	a	Los Angeles
2	Chicago	b	Multnomah
3	Los Angeles	c	Wayne
4	Dallas	d	King
5	Atlanta	e	Harris
6	Oakland	f	Dallas
7	Houston	g	Cook
8	Portland, Oregon	h	Alameda
9	Seattle	i	Dade
10	Detroit	j	Fulton
11	Minneapolis	k	Maricopa
12	Phoenix	l	Hennepin

32. Here are sets of suburbs. Can you name their city?
 a Scottsdale, Mesa, Tempe
 b Chalmette, Metairie, Gretna
 c Millbrae, Sausalito, San Bruno
 d National City, La Jolla, Coronado
 e Aurora, Arvada, Lakewood
 f Sunrise, Plantation, Dania
 g Brookfield, Elmhurst, Cicero
 h Rockville, Bethesda, Falls Church
 i Brookline, Cambridge, Revere

 j Warren, Hamtramck, Grosse Pointe
 k Florrisant, Kirkwood
 l Shaker Heights, Parma
 m Covington, Chevoit, Norwood
 n Gresham, Beaverton, Lake Oswego
 o Upper Darby, King of Prussia
 p Pasadena, Jacinto City, Humble
 q Farmers Branch, Mesquite, Irvine
 r Renton, Burien, Kent
 s Wauwatosa, Whitefish Bay

33. About how many taxis could you park in New York's Central Park if the park were paved?
 a 100,000
 b 300,000
 c 3 million

34. The rock dove, otherwise known as the pigeon, lives in many cities, where one of its activities is the cause of respiratory problems among many of the humans. Name the activity.

35. What and where is Harborplace?

36. In which city does the Bunker Hill battle monument stand?

37. What city is across the Snake River from Lewiston, Idaho?

38. What does James Fenimore Cooper have to do with baseball?

39. The economic core of the eastern United States is the megalopolis that runs from _____ to _____ (pick two):

Southern Maine	Northern Virginia
Southern New Hampshire	Northern Delaware
Southern New York	Northern North Carolina
Southern Maryland	Northern Rhode Island

40. Here is a list of rivers and the cities found on their banks. Imagine you are facing downstream, and tell which bank, left or right, each city or feature is found on:
Manhattan on the Hudson _____
The Jefferson Memorial on the Potomac _____
Saint Louis on the Mississippi _____
New Orleans on the Mississippi _____
Philadelphia on the Delaware _____
Cincinnati on the Ohio _____
Sacramento on the Sacramento _____
Minneapolis on the Mississippi _____

41. Where is the most crowded town in the most crowded county in the most crowded state?

42. In Los Angeles, the neighborhood west of the Wilshire Country Club and north of the La Brea fossil pits is one of luxury apartment buildings. What would you have found in this area in the 1920s?

43. Joplin is the name of a small city in the Midwest, whose main attraction is a mineral museum. People do not drive far to see minerals. At least two other Joplins, however, have made lasting contributions to American culture. Can you name them? (Hint: There's a bit of a trick here.)

44. Do you know the way to San Jose?

U.S.A. Cities

1. Buffalo wings are the contribution of Buffalo, New York, to hot cuisine. They are chicken wings, undipped and uncoated, deep-fried until reddish-brown and served—frequently as happy-hour bar food—in a sauce of melted butter and Tabasco.

2. Boston, New York, Philadelphia, Baltimore, and Charleston were the largest in 1790.

3. 1–f   4–c
   2–b   5–a
   3–e   6–d

4. There are Warsaws in New York, Illinois, Indiana, Kentucky, Missouri, North Carolina, Ohio, and Virginia. There are Troys in New York, Alabama, Kansas, Michigan, Missouri, North Carolina, and Ohio. There are Uticas in Michigan and New York. There are Syracuses in Kansas and New York. There are Versailleses in Connecticut, Indiana, Kentucky, Missouri, and Pennsylvania. There are Parises in Arkansas, Idaho, Illinois, Kentucky, Maine, Missouri, Tennessee, and Texas (and Ontario). There are Cairos in Georgia and Illinois (the latter was "Eden" in Charles Dickens's novel *Martin Chuzzlewit*).
   There are Alexandrias in Indiana, Kentucky, Louisiana, Minnesota, South Dakota, and Virginia (there is also one in Romania). There are Memphises in Missouri, Texas, and Tennessee.

5. 1–i   6–h
   2–g   7–a
   3–d   8–e
   4–f   9–b
   5–c   10–j

6. a Portland, Oregon   f Portland, Maine
   b New York City   g Chicago
   c Houston   h San Diego
   d Philadelphia   i San Francisco
   e Denver   j Saint Louis

7. Oakland, California, her "home town."

8. Dogs drop twenty thousand tons of dog droppings annually in New York.

9. These cities were silver-mining towns in the late 1800s.

10. These stops on your Monopoly board game are all named for real places in Atlantic City, New Jersey.

11. Memphis, the city of Elvis Presley and Johnny Rivers, is on the Mississippi.

12. Cincinnati, the Queen City, the city of Pete Rose and WKRP, is on the Ohio River.

13. El Paso is on the Rio Grande.

14. Omaha, across the river from Council Bluffs, is on the Missouri River.

15. Milwaukee, beer capital of America, is on Lake Michigan.

16. Tacoma is on Puget Sound.

17. San Jose is on San Francisco Bay.

18. Yuma is on the Colorado River.

19. Jefferson City, Missouri; Pierre, South Dakota; and Bismarck, North Dakota, are all on the Missouri River.

20. Henry David Thoreau lived in Concord, Massachusetts. So did his friend Ralph Waldo Emerson. Also in Concord is the Revolutionary battleground where was fired "the shot heard round the world."

21. A meeting in 1852, in Camillus, a small town near Syracuse, led to the formation of the G.O.P.

22.
1–e	4–c
2–f	5–b
3–d	6–a

23. The city of Johnsons and Andersons is Minneapolis, where Scandinavians make up 12 percent of the population.

24. William Penn founded Philadelphia in 1682 as a "holy experiment." From 1790 to 1800 it was the nation's capital, and it had the country's first bank, hospital, and zoo.

25. "Blade 125," if it was graffiti in the New York City subway—and it was—is probably the nickname of someone who lives on One Hundred Twenty-fifth Street and gives a hint as to his or her personality.

26. The original name of Moscow, Idaho, which you might reasonably have thought was Saint Petersburg, or Minsk or Pinsk, was Hog Heaven. Early settlers noticed that their hogs rooted happily for a heavenly prairie weed called camas root, and so they called the place Hog Heaven. Later on, the city fathers decided to give it a more seemly name, and went too far, as people do, calling it Paradise. Paradise Ridge still exists north of town. But when the city was incorporated, they named it Moscow—after a town in Pennsylvania that somebody important had come from. Aren't humans wonderful?

27. The first coin laundry was in Fort Worth, Texas. The Washeteria opened with four washing machines on April 18, 1934.

28.
1–r	7–l	13–g
2–q	8–k	14–f
3–p	9–j	15–e
4–o	10–i	16–c and l
5–n	11–a	17–d
6–m	12–h	18–b

29. L'Enfant Plaza is in Washington, D.C. It is named for Major Pierre Charles L'Enfant (1754–1825), whose design the city is largely based upon. L'Enfant superimposed a network of great diagonal avenues upon a rectangular grid of streets, using circles (Dupont Circle, etc.) to articulate the two systems. Squabbling politicians sacked the designer before construction of the new capital began. The job was done substantially as he had envisioned it—except that monuments to Lincoln and Jefferson were added.

30. Ferris built the world's first Ferris wheel in Chicago, for the World's Fair. It was 264 feet high, higher than the tallest building of the day, and a spectacular hit. People are still building and riding on Ferris wheels.

31.
1–i	7–e
2–g	8–b
3–a	9–d
4–f	10–c
5–j	11–l
6–h	12–k

32. The suburbs given are found surrounding the following cities:
a Phoenix
b New Orleans
c San Francisco
d San Diego
e Denver
f Fort Lauderdale
g Chicago
h Washington, D.C.
i Boston
j Detroit
k Saint Louis
l Cleveland
m Cincinnati
n Portland, Oregon
o Philadelphia
p Houston
q Dallas
r Seattle
s Milwaukee

33. You could park 300,000 taxis in Central Park. But they'd all be stripped in the morning.

34. Like the city dog's, the city pigeon's most unsavory activity is the dropping of droppings. When dried to dust, they are inhaled and cause respiratory problems for many people. Not surprisingly, these medical problems are not always accurately diagnosed.

35. Harborplace is the "restored" Baltimore waterfront—an effort to revive a once successful area whose old function, as a seaport, has evaporated, by turning it into a tourist attraction and shopping center.

36. The monument to the Battle of Bunker Hill is in Boston—on Breed's Hill, where the battle actually took place.

37. Clarkston, Washington, is across from Lewiston, Idaho. The exploring partners passed here on their voyage to the West Coast for President Thomas Jefferson.

38. Cooperstown, New York, was founded by the author's father in 1785, and is the site of the Baseball Hall of Fame.

39. The East's great megalopolis and economic core runs southward from southern New Hampshire to northern Virginia, more than four hundred miles. It is the most urbanized, industrialized segment of the continent, containing New York City, Philadelphia, Boston, Washington, Baltimore, and Newark and Trenton. The eastern edge of this area is well defined by the Atlantic Ocean, but the location of its western edge is debatable. This is the Middle Atlantic region of the United States, and while there still is a great deal of farming done here, as well as a great deal of industrial manufacturing, the region's product is decisions. This is where most of the people live and work who, as one author put it, "serve, educate, direct, finance or govern."

40. Manhattan is on the left bank, the Jefferson Memorial on the left, Saint Louis on the right, New Orleans on the left, Philadelphia on the right, Cincinnati on the right, Sacramento on the left, and Minneapolis on the right.

41. This award goes to West New York, Hudson County, New Jersey. According to the 1980 census, nearly forty-four thousand people live in West New York's area of just under one square mile. West New York accomplishes this feat by severely limiting commercial, industrial, or park lands. There are no empty lots, but there are great views of the Manhattan skyline across the Hudson River.

42. This area of Los Angeles was studded with oil wells in the 1920s.

43. Scott Joplin, of Missouri, gave us the sweet, clear beauty of ragtime music; and Janis Joplin, of Port Arthur, Texas, was one of the first American women with nerve enough to sing real rock and roll, expressing all the romantic, self-important pain and aggression that rock music requires.

44. To get to San Jose from Hollywood, as good a place as any to start out, you take the Hollywood Freeway to the Ventura Freeway, which is U.S. 101. Follow this west and north to San Jose, California. Of course, if the California San Jose is not the one you want, you'll have to follow other directions—and there are many other San Joses in the world, including at least three in Venezuela, six in Mexico, three in Bolivia, one in Colombia, one in Costa Rica, one in Guatemala, one in Peru, four in the Philippines, five in Argentina, two in Panama, one in Uruguay, one in Puerto Rico, one—possibly the first one—in Spain, as well as one in New Mexico and one in Illinois.

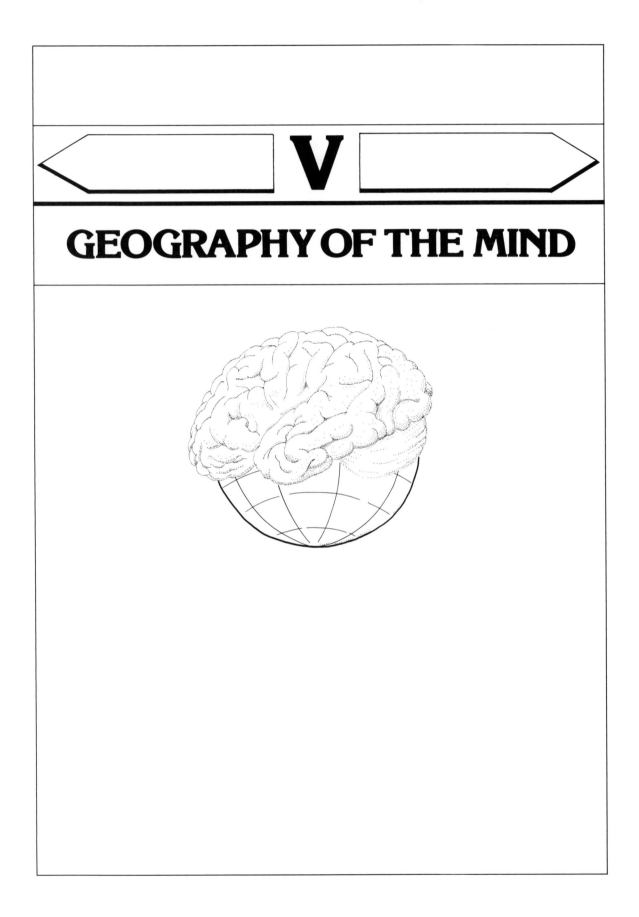

V

GEOGRAPHY OF THE MIND

MAPS OF THE SPIRIT

Some "places" in the "world" have as much to do with feeling, thinking, and perception as with what or where or whose places they actually are. They exist, perhaps, but how they exist is partly in the mind. Remember Saul Steinberg's now-classic *New Yorker* cover, depicting New York City as the center of everything and the rest of the nation as dwindling insignificantly off into the south and west? What made that such wonderful satire was its imaginatively comic yet truthful representation of the feelings of certain provincial Manhattanites. That this same cover has become a framed icon found on provincial walls all over the country makes the ironies all the more delicious.

That's geography of the spirit, you might say. You can find and measure a place like Nepal or Montana, and

even tell pretty much whom it belongs to and who made it what it is. But some places take art to explain.

This is a section of questions about places that exist, if they exist at all, at least partly in the human—or some other—mind, or take their significance from the way they are seen, felt, experienced, or understood by people, or by cats, or by God, or something. We are talking here about the geography of ideas, values, and perception, of territoriality and ownership, of sovereignty and sharing, of aspiration, hope, and superstition, of "real" space and symbolic space.

This, in human terms, is probably the most fascinating aspect of geography. It could fill another book. Here is a sample:

1. In what country was the first Christian state?

2. Where in the world will you find love? Romantic love, that is, in the *Cosmo Girl* sense. Where are its boundaries?

3. There are five spots on the globe that most people would probably describe as ideal to live in. They have what is called a Mediterranean climate, and the edge of the Mediterranean sea—except for northern Egypt and Libya—is a good example. This climate is characterized by moderation, by westerly

winds prevailing across cooling ocean currents, and by low rainfall—averaging twenty inches a year or less—mostly in the winter when the sun is low. In the summers, the sun is hot, but there is little rain, so it is not humid. Despite occasional fogs, the weather is crisp and dry most of the year. All these wonderful places are on or near coasts. As we've said, the Mediterranean is one of them. Where would you find the others? Hint: An earlier section gave away one answer.

4. What determines a suburban cat's territory?

5. Match the traditional enemies:

Cowboys	Turks
Greeks	Cambodians
Russians	Israelis
Arabs	Indians
Vietnamese	Germans
Japanese	Chinese

6. What denomination was the world's first drive-in church? Extra points if you can say where and when it was founded.
Holy Roller
Baptist
Presbyterian
Roman Catholic
Lutheran
Methodist
African Methodist Episcopal

7. Where was the first ghetto?

8. By simply observing statistics, can you predict which country may be ripe for a modern violent revolution? If so, how?

9. Where on earth are you most likely to survive a nuclear holocaust?

10. And what is the weight of a scruple?

11. Here is a pretty insightful quotation: "The great nations have always acted like gangsters, and the small nations like prostitutes." What sort of person would—and in fact did—say a thing like that? Extra credit if you can name him or her.
A politician
A filmmaker
A philosopher
A comedian
A stand-up geographer

12. Which does not belong on this list?
Xanadu
Dildo Key
Atlantis
Never-Never Land
Shangri-La
Land of Cockaigne

13. Where is the largest fenced-in area in the world, and what is it used for?

14. Match the country with its former colonial ruler:

1	Haiti	a	Italy
2	Guyana	b	U.S.A.
3	Indonesia	c	United Kingdom
4	Vietnam	d	Denmark
5	Ecuador	e	Belgium
6	Angola	f	Portugal
7	Brazil	g	Spain
8	Libya	h	France
9	Zaire	i	Netherlands
10	Belize		
11	Greenland		
12	Philippines		

15. Match these landforms with the war in which they were prominent:

1	Yalu River	a	World War I
2	Camranh Bay	b	Mexican-American War
3	San Juan Hill	c	Civil War
4	Omaha Beach	d	American Revolution
5	Bunker Hill	e	World War II
6	Prairie Grove	f	Spanish-American War
7	Cerro Gordo	g	Vietnam War
8	Somme River	h	Korean War

16. Early Chinese immigrants to the United States called America *Gum Shan.* What did this mean?

17. The word *California* comes from early Spanish folklore. What does it mean?

18. Where were Sodom and Gomorrah?

19. Match the country with its state or dominant religion:

1	United Kingdom	a	Roman Catholic
2	Sweden	b	Lutheran
3	Italy	c	Anglican
4	Greece	d	Greek Orthodox
5	Turkey	e	Islam
6	Australia	f	Beer
7	Indonesia	g	Dutch Reformed
8	South Africa	h	Buddhist
9	Haiti	i	Jewish
10	Thailand	j	Christian/voodoo
11	Israel		
12	Uruguay		

20. Roman Catholicism is to Saint Peter's Basilica what the Church of England is to _____ _____.

21. Zoroastrianism is the ancient religion of the Parsi, the people of ancient Persia. Which city of the world now has the largest Parsi community?

22. Six hundred million humans are Muslims. There are five pillars of Islam. One of them is a mandatory pilgrimage, at least once in a lifetime, to where? Name the city—and its country.

23. Perhaps the most common religion in East Asia and Southeast Asia is Buddhism. Where was the Buddha born?

24. Tom Shales, a Washington, D.C., television critic, says, "Bad news, the real news of the day, comes mostly from the East Coast." Where does he say the good news comes from?

25. What do Mizaru, Kikazaru, and Iwazaru—the Buddhist monkeys—represent?

26. Complete this Chinese saying: "In the sky there is Paradise; on earth there is _____."

27. There is a principle in journalism often known as Afghanistanism. Ponder this admittedly obscure statement, apply it to humans and geography, and then complete this sentence: "If you think a thing is pretty weird or unsavory, that thing is probably from _____ _____. The rough idea will do. Hint: Why is the same food called both mountain oysters and prairie oysters?

28. Where did Plato place Atlantis?

29. What is Paradise? Boccaccio said it is (fill in the blanks if you can) "a whole mountain of _____ _____ . . . on top of which stand people doing nothing but making _____ and _____."

30. In the movie *My Dinner with Andre*, Andre kept talking about this paradisiacal commune he visited in a place called Findhorn. Where is Findhorn, really?

31. Residents of which U.S. city consume the most prune juice per capita?

MAPS OF THE SPIRIT

1. Legend has it that the early Christian evangelist Gregory the Illuminator converted King Tiridates III to the Christian faith in A.D. 301. King Tiridates had been turned into a pig by demons, and Gregory drove them out, so in gratitude Tiridates converted his entire kingdom of Armenia to Christianity. What remains of ancient Armenia is now part of the USSR—an atheistic state.

2. The boundaries of romantic love are pretty much identical with those of the developed Western world, which is the only region, by and large, that can afford it, according to the late Peter Farb. People in other places have usually married or otherwise allied themselves for more practical reasons.

3. The Mediterranean climate can be found in four other spots on the globe besides the rim of the Mediterranean. They are in South Africa near Cape Town, in central Chile, along the California coast, and in southwestern and southern Australia near Adelaide.

4. A suburban cat's territory coincides with the yard of the family that feeds it. Until someone else starts feeding it. Of course cats—especially toms—range beyond their own yard, and they don't know as precisely as their owners where their yard ends. But they have a pretty good idea. Most cat fights are on territorial borders.

5. You see how complicated it is! The answers we had in mind are cowboys and Indians, Greeks and Turks, Russians and Germans, Arabs and Israelis, Vietnamese and Cambodians, Japanese and Chinese. But these are not the only answers, as you no doubt

noted. The Russians and Israelis, the Russians and Chinese, the Vietnamese and Chinese, and for all we know, the cowboys and Cambodians, at one time or another have been at each other's throats.

6. The first ecclesiastical drive-in was the Whitfield Presbyterian Church, established in 1951 on U.S. 41, outside Bradenton, Florida. There had been church services held in drive-in movie theaters before, but this was the first designed and built specifically for car-bound worshipers.

7. The ghetto, by one name or another, meaning the segregation of a particular group, has probably existed as long as humans have. The first to have the name was in Venice, beginning in the year 1513, when Jews were cordoned into a neighborhood known as Geto, the site of a former foundry. Thus the word *ghetto* comes from the Italian.

8. Apparently, you can predict revolution from a single statistic. The one to watch is the percentage of landless peasants among the population: When it reaches 40 percent, people begin beating their plowshares into swords, looking for a share. Mexico's revolution occurred when 60 percent of the people were landless, but people were more patient in 1911. In Russia, it was 40 percent in 1917, in Spain it was 40 percent in 1936, in China it was 40 percent in 1945, in Cuba it was 39 percent (Cubans tend to be volatile) in 1959, in Nicaragua it was 40 percent in 1979.

9. Maybe nowhere. But New Caledonia, in the southwestern Pacific Ocean, has the best prevailing winds and currents to prevent radioactivity drifting in

from most-likely-to-be-heavily-hit spots, and its soil is good for self-sufficient agriculture. On the other hand, this overseas territory of France is experiencing an increase in domestic violence recently, between colonials, the Tonkinese (who were originally brought here as field hands), and the natives, called Kanaks.

10. Scruples of the moral sort weigh heavier on some people than on others—as if you didn't know. There's another sort of scruple, however: a unit of weight used by apothecaries, and equal to twenty grains. A grain—taken from the weight of a grain of wheat—is the smallest unit of weight measurement used in the United Kingdom and United States; there are seven thousand of them in a pound.

11. A filmmaker, Stanley Kubrick, who made *A Clockwork Orange*, said it. The one thing a filmmaker cannot be is blind.

12. Dildo Key does not belong on this list because it actually exists, in Florida Bay. The other places are imaginary. But there is a Paradise Island in the Bahamas and a Paradise Valley in Arizona and a town called Paradise in northern California.

13. The largest fenced area is Etosha National Park, a paradise of African wildlife in Namibia (South-West Africa). Its five-hundred-mile fence encloses 8,598 square miles of territory.

14.
1–h	7–f
2–c	8–a
3–i	9–e
4–h	10–c
5–g	11–d
6–f	12–g and b

15.
1–h	5–d
2–g	6–c
3–f	7–b
4–e	8–a

16. *Gum Shan* means "Mountain of Gold."

17. *California* means "earthly paradise."

18. Nobody knows exactly, since anyone who could say for certain is dead or turned to salt, but the best current guess is on the southwestern side of the Dead Sea Valley. The Dead Sea is called dead because it contains a Lot, so to speak, of salt.

19.
1–c	7–e
2–b	8–g
3–a	9–j
4–d	10–h
5–e	11–i
6–f	12–a

20. Canterbury Cathedral.

21. There are eighty thousand Zoroastrians in Bombay, to which many fled following the Islamic conquests of the eighth century. In recent years, Iran's Zoroastrians have been persecuted by Khomeini's fundamentalists.

22. Mecca, in western Saudi Arabia. A small, square building of black stone in the center of the Sacred Mosque in Mecca is a kind of distillation of Islamic holiness. It is called the Ka'aba, the House of God, and the Muslim pilgrim shows his reverence by walking around it seven times counterclockwise.

23. Siddhartha Gautama, the Buddha, was born around 563 B.C. in the Himalayan foothills of Nepal.

24. Tom Shales says, "The good news, the entertainment TV, comes from the West Coast, where hope does handsprings eternal."

25. Mizaru, Kikazaru, and Iwazaru are the names of the three monkeys that, in the Tendei sect of Buddhists, manifest the Three Truths: No See, No Hear, No Speak.

26. "In the sky there is Paradise; on earth there is Hangzhou," is what the Chinese say. Hangzhou used to be called Hangchow, an unlovely name for such a lovely small city built on the shores of natural West Lake. Hangzhou is a vacation spot for lucky Chinese people.

27. Prairie oysters, as it was once explained to one of this book's co-authors by a mountain dweller, are sheep testicles, eaten only by people who live on the prairie, hence their name. However, these same delicacies, according to a prairie dweller, are properly called mountain oysters, since they are consumed only by mountain dwellers, and people who live on the prairie would never touch such fare. The weird and unsavory always comes from *somewhere else*. Of course, they both ate them. And what is Afghanistanism? It is the journalist's habit of managing to uncover the worst misfeasance in places far from home—where the repercussions are more convenient to deal with.

28. Plato figured Atlantis was situated beyond the Pillars of Hercules (the Straits of Gibraltar). It was, he thought, larger than Asia Minor. Legend has it that Atlantis sank and is marked only by shoals. Nobody believes this anymore except Erik von Danniken.

29. The Boccaccio quotation defines paradise as "a whole mountain of Parmigiano cheese . . . on top of which stand people doing nothing but making macaroni and ravioli."

30. Scotland.

31. Miami, Florida.